VEDIC
METEOROLOGY
(Vedic Science of Weather Modification)

(Third Revised and Updated Edition)

By

Dr. Ravi Prakash Arya

सर्वज्ञानमयो हि वेदः

Amazon Books, USA

in association with

Indian Foundation for Vedic Science
1051, Sector-1, Rohtak, Haryana, India Pin124001
Ph. No. 09313033917; 09650183260
Email: vedicscience@hotmail.com; vedicscience@rediffmail.com
Web: www.vedicscience.net

Third Edition

Kali era: 5119 (c. 2017)
Kalpa era: 1,97,29,49,119
Brahma era: 15,55,21,97,29,49,119

ISBN 81- 87710-17-9

Dedicated

Ram Narain Arya (1936-2010)
A perfect Rainmaker of the Modern age

Abbreviations

AB.	*Aitareya Brāhmaṇa*
Aeolus 1952	*Meteorology*, English University Press Ltd. London
AV.	*Atharvaveda*
Āp. Śr. Sū	*Āpastambe Śrauta Sūtra*
Āśv. Śr. Śu	*Āśvalāyana Śrauta Sūtra*
BD., Bṛd, Bṛhd.	*Bṛhaddevatā*
BS.	*Bṛhatsaṁhitā*
BU.	*Bṛhadāraṇyakopaniṣad*
Ch.U.	*Chāndyogya Upaniṣad*
CS.	*Caraka Saṁhitā*
Go.Br.	*Gopatha Brāhmaṇa*
JB.	*Jaiminiya Brāhmaṇa*
Kau. Br.	*Kauṣitaki Brāhmaṇa*
Kāt. S	*Kāṭhaka Saṁhitā*
Kāṇ. Ś. Br.	*Kāṇva Śatapatha Brāhmaṇa*
KPS.	*Kapiṣṭhala Saṁhitā*
K.S.	*Kaṭha Saṁhitā*
M.S.	*Maitrāyaṇī Saṁhitā*
Nir.	*Nirukta*
PM.	*Patañjali's Mahābhāṣya*
R.V.	*Ṛgveda*
Ś.Br.	*Śatapatha Brāhmaṇa*
S.K.	*Sāṁkhyakārikā*
SS.	*Sāṁkhya Sūtra*
SV.	*Sāmaveda*
TĀ.	*Taittirīya Saṁhitā*
Tait. Up., TU.	*Taittirīya Upaniṣad*
Tāṇ. Br.	*Tāṇḍya Brāhmaṇa*
TB.	*Taittirīya Brāhmaṇa*
TS.	*Taittirīya Saṁhitā*
VP.	*Viṣṇu Purāṇa*,Geeta Press, Gorakhpur Saṁ. 2041.
VS.	*Vājasaneyī Saṁhitā*
YD.	*Yoga Darśana*

Contents

Ram Narain Arya performing rainmaking yajña

Preface

The advance of science and technology in modern times has revolutionised the human life in all spheres. This revolution, however, like all other revolutions has backfired. While at one end it has made the life fast, easy and comfortable, it has at the other end created the horrors like an ecological crisis, global pollution and depletion of the ozone layer.

Under the circumstances, the science of meteorology seems to be the only remedy for all the deadly problems. In fact, it is the only panacea for the environmental scourges that are horrifying humanity the world over. As such, the science of meteorology has the most significant and vital role to play in the sustenance of organic life on the planet.

In view of the above fact, several meteorological organisations, World Meteorological Organisation being prominent of all, have come up in various parts of the globe. All these organisations are too incompatible to meet the challenge before us. The modern experimental meteorology is able to invent some possible solutions for the ecological problems that demand immediate attention.

Thus in the present context when modern science is failing, it becomes imperative to excavate the old treasure of science of Vedic Indians for some panacea to the ecological crisis that we are facing today.

The Vedic seers were the masters of vision. They visualised the secret laws of nature beyond time and space. Though they too exploited the maximum natural resources for the organic life to flourish beautifully, they were fully aware of their duty to safeguard the planet and her environment. They didn't destroy it like us, but they kept it intact. They enterprised for natural harmony all through heaven (celestial sphere) and earth (terrestrial sphere). They cherished for placid and harmonised stars, mid-sphere, planets and the planetary flora. Accordingly, they prayed for as under:

dyau śāntir antarikṣaṁ śāntiḥ pṛthivi śāntir
āpaḥ śāntir oṣadhaya śāntiḥ vanaspatayaḥ
śāntiḥ viśvedevāḥ śāntir brahma śāntiḥ sarvaṁ
śāntiḥ śāntireva śāntiḥ sā mā śāntiredhi

VS. 36.17

May there be peace in the light space, peace in the intermediate space, peace in the observer space. May the waters flow peacefully. May the herbs and shrubs, with powers of healing, grow peacefully. May all natural forces bring us peace. Knowledge grants us peace. Everything should bring us peace and peace be peace in the true sense. Let that peace come to me.

The present work is a humble attempt towards supplementing the modern meteorology with a new chapter, what may be called an 'old', on the innovations and inventions of Vedic *Ṛṣis* in this sphere.

The present work consists of two parts. The first part being 'Meteorology', deals with the meteorological phenomena such as vaporization, the formation of clouds, induction of an electrical charge in the clouds, i.e. lightning, thunderstorms, precipitation of raindrops, hailstones and snowflakes as visualised and defined by Vedic seers.

It presents for the first time an account of the factors that form the meteorological phenomena on the earth and also the factors that were considered by the Vedic Meteorologists as having control over the weather conditions on the earth.

It also deals with the means of short-range and long-range forecasts adopted by ancient Indians from time to time.

The second part consists of Experimental Meteorology. It is an elaboration on as to how to fetch the celestial river Ganga on the earth as Bhāgiratha did at a time when the planet was ridden with water crises. The Ṛgvedic seer suggests the local inhabitants cut the celestial waters down the earth not with the help of material means like spades, daggers and other weapons, but with the help of oblations offered to *yajṇīya* fire.

For the first time, it tries to lay down the theory and principles of Vedic Science of *Yajña*. It briefs up readers with the efforts of rainmaking put in by various scholars from time to time in the history of mankind and finally takes stock of the attempts of rain making with the help of *Yajña* from the Vedic period down to Ram Narain Arya who has performed over 150 successful experiments on rain, anti-rain, change of the flow and direction of winds and prevention of diseases and air pollution in various parts of this country and thus helped promote the Vedic Science for the well-being of mankind. It also accounts for the methods of rainmaking as enshrined in the Vedas and tried upon by the rainmaker during the course of his experiments on rainmaking.

It throws open a new chapter for research on the prevention of rain (anti-rain) for the scientists of experimental meteorology. So far the meteorologists have made attempts of rainmaking to avoid famines or scarcity of water, but none seems to have attempted the experiments on prevention of falling rain to overcome the problems of floods or excess of water.

In this portion, the curious readers/ researchers will find a concrete and scientific basis behind such mythological descriptions of Indian history as appeared to both learned and laity nothing else but fabulous, miraculous and other fallacious speculations of our elders. So far, the readers have learnt from the columns of history and mythology that rain could be induced or stopped by shooting a particular type of arrows in ancient times. But, here, they will find that they can also do what was earlier often talked about by way of references to history and mythology.

The thrilling idea of rainmaking or anti-rain with the help of *Yajña* is not only cherished by the present author and the subject rainmaker, but modern scientists have also reached to some nonetheless thrilling conclusions, such as:

(1) The better effectiveness of silver iodide as a cloud seeding material when introduced in the sky in the form of smoke from the ground with the help of surface burners or ground generators than when delivered by airplane.

(2) Burning sugarcane in eastern Australia is thought to have reduced precipitation near the burning areas through overseeding.

These conclusions of modern meteorologists also validate the authenticity of the idea of rainmaking and anti-rain with the help of *Yajña.*

To sum up, it can unhesitatingly and safely be said that the Vedic meteorologists studied the nature up to the minutest detail possible and modified weather effectively as per their needs and requirements with the help of the technique of *Yajña* based on concrete and scientific ideas which need to be revived and revitalized. Through these lines, the author would like to invite the attention of the world Government and scientific bodies that they should come forward to save the scientific treasure of the Vedas from being threatened by extinction and preserve, popularize and promote this ancient tradition of *Yajña* by establishing *Yajña Saṁsthānas.*

In the end, I must say that we are happy to bring out this revised and updated second edition of Vedic Meteorology before the readers. Since the time of publication of the first edition several new types of research have come out. The results of all those new findings have been incorporated into this edition in the form of appendices. For example appendices III and IV have been added on '*Agniṣomīya Paśuyāga*-A Vedic operation of Rainmaking' and '*Somayāga*-A Process of rainformation'. Some of the contents forming the chapters of the first edition which were found not so relevant as to be dealt within *Vedic Meteorology* have also been deleted. For instance, Appendix III on '*Nature of Science and Technology in the Vedic Age*' and Chapter 2 on '*Origin and Evolution of Universe*' have been deleted from this edition as separate books have been demanded by our worthy readers on these two subjects. Hope the readers will find the revised and updated second

Dr. Ravi Prakash Arya

1

Meteorological Phenomena

The visible universe or the physical world is composed of five gross elements *viz. ākāśa, vāyu, agni, āpaḥ* and *pṛthivi*. The *ākāśa* provides space and the *pṛthivi* provides the material base for the operation of the universe. the rest of the elements *vāyu* (air), *agni* (fire), *āpaḥ* (water) play their significant roles in the operation of the universe. Actually, the whole atmosphere of the earth is composed variously of air, fire and water. Hence, the meteorological phenomena on the planet of the earth are formed mainly of air, fire and water and their variables. For instance, the origin of winds, storms and hurricanes etc. may be assigned to the air; the cyclones and the phenomena of hotness and dryness in the atmosphere may be studied as the variables of the fire; the clouds, lightning, thunder or precipitation (drizzle drops, raindrops, snowflakes and hailstones) may be studied in the name of water-variables.

catvāri vā apāṁ rupāṇi megho vidyut stanayitnur vṛṣṭiḥ.[1]

In addition, several other physical forces popularly known as deities play their vital roles in the materialisation of meteorological phenomena on the earth in a variety of manners. For example, *marutas* provide lightning with flash,[2] hence help in its occurrence. They also cause the thunder (*aśani*) by inducing electric charge[2] into *āpaḥ*.

maruto' dbhiragnim atanyan tasya tāntasya hṛdayam ācchindan so aśnir abhavat.[3]

'The *marutas* induce electric charge into *āpaḥ* i.e. vapours abiding in the midspace. The (-ve) electric charge induces its (+ve) charge and gets neutralised turning into thunder.

tasya (agneḥ) marutaḥ stanayitnunā hṛdayam ācchindam sa divyāśnir abhavat.[4]

'The *marutas* shattered the nucleus of the fire/*vidyut* which resulted in the thunder-bolt.'

The *marutas* also control the electricity in the midsphere, that helps form water.

asthāpayanta yuvatiṁ yuvānāḥ śubhe nimiślāṁ vidatheṣu prajām.[5]

The *marutas* also stimulate clouds to yield rain.

marudbhiḥ pracyutā meghā varṣantu pṛthivīmanu.[6]
marudbhiḥ pracyutā meghā saṁyantu pṛthivīmanu.[7]

Marutas also cause the vaporisation of oceanic waters into midsphere and the sun stirs the water vapours over and above in the midsphere and helps form the clouds.

udirayat maruraḥ samudratastveṣo arko
nabha uta pāthyāth mahaṛṣabhasya nadato
nabhasvato vāśrā āpaḥ pṛthiviṁ tarpayantu.[8]

See also

prajāpatiḥ salilādā samudrādāpa
irayannuddhimardayati prapyāyatāṁ.
vṛṣṇo aśvasya reto' rvāṅtena stanayitnunehi.[9]

The Vedic seers also had the vision that midsphere was the main centre of operation of air, fire and water and their variables often called as deities for creating the meteorological phenomena on the earth.[10]

According to them:

vāyur vā antarikṣyādhyakṣaḥ.[11]

'*Yāyu* is the dominating occupant of midsphere'

vāyur asyantarikṣe śritah.[12]

'*Vayu* has got the shelter in midsphere.'

(*vāyur antarikṣe* (*dīpyate*).[13]

'*Vayu* shines in the midsphere.'

idam eva āntarikṣaṁ jyotiḥ.[14]

'The fire has also housed itself in the mid-space.'

apāmupasthe apāṁ sthāne antarikṣe.[15]

'Midsphere is the centre of water-vapours.'

Yāska, a leading Vedic Scholar of ancient India, has also assigned the territorial division of midsphere to all such deities/physical forces as the effect the atmosphere or climatic conditions on the earth. According to him, '*Indra*' and '*Vāyu*' govern the midsphere as presiding deities and *Parjanya, Rudra, Bṛhaspati, Apāmnapāt, Pururavā, Aditiḥ, Tvaṣṭā, Savitā, Vāta, Vācaspati, Soma, Marut, Aṅgirasa, Ṛbhu, Pitara,* etc. dominate the midsphere as co-deities or subordinate deities.[16]

This is why the Vedic seers studied the meteorological phenomena in the name of *antarikṣa/dyau*[17] i.e. midspherical-phenomena.

Thus from the foregoing discussion, we may infer that the Indian meteorological concept dates back to the age of *RV.* Itself. Later Vedic *Saṁhitas* and still later Vedic literature are also abound in meteorological observations. This makes it evidently clear that this branch of science was developed to the fullest extent in the Vedic period itself. In the later Vedic and post-Vedic period, we come across such sages as Garga, Parāśara, Kāśyapa, Ṛṣiputra, Vajra,

Bādrāyaṇa, Asita, Devala, etc. who produced the works of book-length on meteorology. Though their actual works have been lost in the hoary past, we are able to gather fragmentary references to their meteorological concepts and observations from Varāhamihira's famous astrological treatise *Bṛhatsaṁhitā*. For instance

> *tallakṣaṇāni munibhiryāni nibaddhāni tāni dṛṣṭvedaṁ kriyate Garga Parāśar-Kāśyapa-Vajrādi racitāni,*[18]

'Keeping in view the symptoms of meteorological phenomena as have been explained by the sages like Garga, Prāśara, Kāśyapa and Vajra, I (Varahmihira) intend to compose this work.'

Now, we shall discuss hereunder in detail the part played by water, air, fire and other physical forces operating in midsphere for determining weather conditions/climatic conditions or meteorological phenomena on our planet, the earth.

Notes and References

1. *TĀ.* 1.24.1.
2. *Vidyunmahasaḥ. RV.* 5.54.13.
3. *TB.* 1.1.3.12
4. *KPS.* 6.7
5. *RV.* 1.167.6
6. *AV.* 4.15.7
7. *Ibid.* 4.15.8
8. *Ibid.* 4.15.5
9. *RV.* 583.6; *AV.* 4.15.11
10. See for detail Author (1993:14)
11. *MS.* 4.1.1
12. *JB.* 3.2.11
13. *JB.* 1.292

14. *Ibid.* 2.166

15. *Nir.* 7.27

16. See author (1993:14-15)

17. This is to inform here that *dyuloka* (celestial sphere) has though been differentiated from *antarikṣa loka* (midsphere) so far as the territorial division of gaseous and plasmatic matters/masses is concerned, yet in normal course objects abiding in the midsphere (*antarikṣastha vastu)* are also called as *daivī* or *divya,* i.e. the objects residing in the celestial sphere. It is also clear from the statement of *Aitareya Brāhmaṇa* (3.66) *dyau antarikṣe pratiṣṭhitaḥ* i.e. in the normal course of saying *dyau* (celestial sphere) is said to be situated in the *antarikṣa* (midsphere).

2

Jala (Water)

Water is available in the atmosphere in three states, e.g. solid as ice, liquid as water and gas as vapours. As solid, it finds the solid mass of continents i.e. mountains as its abode. As a liquid, it joins its level of liquid mass through flowing down the oceans inland seas. In its gaseous state, it occupies its place in the gaseous mass abiding in the mid-sphere.

The water is one of the main factors that account for weather or meteorological phenomena on the earth. Though its contribution to climate formation in its solid-state as ice and liquid state as water cannot be ruled out, yet it plays a significant role in creating the meteorological phenomena on the earth only in its gaseous state. This is why, the waters in a gaseous state or water vapours, more often than not denoted by the term *āpaḥ,* were the feature of central interest for the Vedic meteorologists.

***Āpaḥ-Variable of Water*:** The term *āpaḥ,* as mentioned above, was coined to signify the gaseous waters or vapours that are transferred from earth's surface, primarily the oceans into mid-sphere. According to the *Ātharvaṇa* seer, *'āpaḥ* are the waters that are transferred from oceans into the mid-sphere.'[1]

Actually, the word *āpaḥ* was derived from the root √*āp* 'to receive or obtain' in view of the concept of its being received by the *Indra* (electrical charge) in the mid-sphere after its vaporization by *Varuṇa* deity (i.e. radiation heating from the sun) from earth's surface. This view may very well be shared with another *Ātharvaṇa* seer. He observes as follows:

> 'On being sent forth (vaporized) by *Varuṇa* from the earth, you receive *Indra* (electrical charge). So, you are called *āpaḥ*.[2]

It may also be noted here that the process of conversion of liquid waters into *āpaḥ* (gaseous waters/vapours) is accomplished by *Varuṇa* deity, co-deity, or subordinate deity of the presiding deity, the *Sūrya.*

In fact, the *Varuṇa* deity is a physical force representing the radiation of heat from the sun. The waters vaporized through radiation heating from the sun give rise to the seeding of *Āgneya* clouds (a detailed break-up of this type of clouds will be given in the following chapter) which become electrically charged and so turn into thunder-clouds. On the other hand, the clouds that are seeded on account of seasonal sea-bearing wind's (like monsoons in India), or on account of moon's conjunction with some particular constellations, don't produce lightning and thunder. This is why, the Vedic seer talks about the waters vaporized through radiation heating as being possessed of *Indra,* the electrical charge. This vaporization of waters due to heat leads to their expansion in terms of volume. The Ṛgvedic *Ṛṣi* points out this expanding characteristic of *Agni* (fire in case of terrestrial *Agni* and heat in case of celestial *Agni*). The verse reads like this:

> 'Let me offer oblations to the fire, which expands *āhutis* offered to it.'[3]

This increase in the volume of liquid substances on being changed into the gaseous state due to heating has been calculated by modern scientists (particularly in the context of the transformation of liquid waters into vapours) by a factor of nearly 1000.[4]

Āpaḥ Cycle / Hydrological Cycle: The liquid waters thus transformed into the midsphere in the form of *āpaḥ* or moisture/vapours cycle back to the earth in their original state.[5] To complete this cycle, they have to pass through four phases. These phases have been enumerated by the *Taittirīya* seer as:

> 'Moisture/vapour undergoes the change of phase as cloud, electrical charge, *stanayitnu* (lightning and thunder) and *vṛṣṭiḥ* (precipitation).'[6]

The first change of phase takes place in the form of clouds. That is, the vapours condensed into clouds. The second change of phase has been observed in the production of electrical charge (*vidyut*) in the clouds. That third change of phase has been studied in the name of production of discharge/spark between the oppositely charged parts of a cloud, or of two neighbouring clouds or between a charged cloud and the neutral earth. The discharge/spark thus produced neutralises the charges (more or less) while turning the electrical energy into light, heat and sound in the spark. The spark so produced is apparent as lightning, while the sound and its reflection from clouds, etc. gives rise to what is known as thunder. The fourth stage of change is described as the reverse change of phase under which the water-vapours again turn into water and fall back to the earth from clouds.

The hydrological cycle is thus the evaporation-precipitation-run off relation. In other words, it consists in the evaporation of waters from the oceans and their falling back on earth in the form of precipitation.

***Fetching the River Ganga*:** Actually, the phenomenon of water's cycling through midsphere back to earth was considered to be of utmost importance to maintain the hydrological balance of the earth. So, to overcome the problem of water shortage, artificial means were also applied to stimulate the *āpaḥ,* i.e. water vapours to fall back on the earth. One of such attempts (the artificial stimulation of rain) has been recorded in the columns of history when the whole earth was afflicted by the problem of water shortage. History has it that to remove the problem of water shortage, the king *Bhāgiratha* fetched the celestial river

Gangā nadī on the earth. In fact, the term *nadī* in the Vedic periods was used for the celestial waters only. This can very well be understood in the light of following observations of the *Ātharvaṇa* seer. According to him, '*nadī* (river) is the name of water-vapours contained in the clouds, which produce the nadan (thundering) sound due to the passing of discharge/spark. This celestial river (*nadī*) transforms into terrestrial one (*sindhu*) after falling from heaven (midshpere) in the form of rain-showers.'[7]

Later on, the same term (*nadī*) came to be applied to the waters flowing on the earth in the form of rivers (*sindhu--syandanā imāḥ bhavanti*) and rivulets. Yāska seems to etymologize this word in the context of terrestrial waters. According to him, 'The terrestrial rivers are called *nadī*, since they produce *nadan* sound while flowing down.'[8]

Thus it is crystal clear from the foregoing illustrations and discussions that the word *nadī* was applied in the Vedic times for the celestial waters contained in the thunder-clouds. This was, actually, the basic concept of the *Gangā nadī* existing in the celestial sphere (*dyuloka/ svargaloka/ suraloka*), or so-called heaven. The *Ṛgvedic* seer, in one of the verses, suggests the local inhabitants cut the celestial waters down to the earth, not through the material means like spades, daggers and other weapons, but through oblations offered to the *Yajñiya* fire in order to irrigate the earth for her inhabitants with waters. The verse reads as under:

> 'Let me offer oblations to the celestial waters (gaseous waters/water-vapours), which are drawn by the sun-rays. This oblation is offered in order to sever them from heaven (midsphere) down to flow on the earth in the form of rivers and rivulets.'[9]

Notes and References

1. *AV.* 4.27.4.

2. *yat preṣitā varuṇenācchibhaṁ samvalgata*
   tadāpnodindro vo yatistasmādāpo anuṣṭhan. *AV.* 3.13.2.

3. *agnim̐ yajadhvam̐ haviṣā tanā girā.* See author (1993: 22)

4. *Mason* (1978: 77)

5. *āpaḥ samudrādivam̐ divaḥ pṛthivim adhi ye sṛjanti.* *AV.* 4.27.4.

6. *catvāri vā apām̐ rupāṇi megho vidyut stanayitnurvṛṣṭiḥ.* *TĀ.* 1.24.1.

7. *yadadaḥ sam̐prayātīrahāvanadatāhate tasmādā nadyo3 nā mastha tā vo nāmāni sindhavaḥ.* *AV.* 3.13.1.

8. *nadyaḥ kasmāt nadanāḥ imā bhavanti śabdavatyaḥ.* *Nir.* 2.25.

9. *apām̐ devirūpam̐ hvaye yatra gāvaḥ pibanti naḥ sindhubhya kartve haviḥ* *RV.* 1.23.18.

3

Megha (Clouds)

As it has already been pointed out above that the *āpaḥ* cycle starts with the first Phase of change as clouds. In which the vapours condense and form the clouds. To elucidate the process of cloud-formation, the author would like to reproduce the Vedic theory (that has already been expounded)[1] of the 'Physical Embodiment of the Universe'. The author has explained thereby the principle of '*yad aṇḍe tad brahmāṇḍe*', 'whatever is contained in an individual physical body is also contained in the physical world or universe.' The same theory finds its applicability in the context of cloud-formation and the precipitation. The physical process involved in the cloud formation and precipitation has been compared by the Vedic sages with the biological process of conception and delivery. The *RV.* 1.6.4 contains a reference to the formation of rain embryos.[2] Sāyaṇa has aptly interpreted this concept. Varāhamihira, a leading astrologer of ancient India, describes the phenomenon of the formation of clouds by the term *garbha dhāraṇa* (conception) and the rainfall/precipitation by the term *prasava* (delivery). Thus while comparing the physical process of cloud formation to the biological process of conception and precipitation to delivery, the present author takes the liberty to go a step further to compare the physical phenomenon of discharge (thunder + lightning) leading to precipitation to the biological phenomenon of labour pains leading to delivery.

Cloud Formation

Clouds are generally formed of water vapours existing in the midsphere. The Vedic meteorologists after a careful and minute examination of nature investigated all these factors that caused the transportation of the terrestrial waters into the mid-space.

Extra-Planetary Factors leading to Cloud-formation

Three extra-planetary factors that work behind the transportation of terrestrial waters in the midsphere were investigated by the Vedic meteorologists.

Celestial Fire (Sun)

The first factor to make the terrestrial waters rise into the midsphere is the celestial fire or the sun. Vedic sages observe this fact as follows:

> '*Agni* (fire) stirs water-vapours from here (Earth).'[3]

The rise of terrestrial waters into mid-space by the sun is popularly known as evaporation. The sun radiates light and heat on the earth. The radiation heating from the sun performs the work required for the transportation of terrestrial waters into midsphere, or say, for the volume expansion from water to vapours. In this evaporation process, or the transportation of waters from the earth to midsphere, the radiation heating from the sun may be only a necessary but by no means a sufficient criterion for evaporation to occur. The air or wind-speed also plays a decisive role for evaporation to take place. This is why, the Vedic seer takes into account the assistance of air for evaporation of waters by the sun. According to him: "The sun assisted by air evaporates the terrestrial waters into the celestial ones and makes them fall on the earth in the form of rain."[4] The similar observation is made by Śaunaka in his *Bṛhaddevatā.* According to him, 'Since the sun in conjunction with air conducts evaporation of waters and makes them fall on the earth, it is known as *Indra* (electric charge) in the midshpere.'[5]

In fact, waters that are evaporated by the radiation heating from the sun (*Varuṇa* deity) are said to have possessed of *Indra*[6], the electrical charge. So, the clouds formed of these vapours become electrically charged. They are the precursor of thunderstorms or they often yield rain followed by lightning and thunder. Hence, they are also known as thunder clouds. Since they originate from fire, they are classified as *Agnija.*

The artificial cloud seeding through *Yajña* often conducted by the Rainmaker, Ram Narain Arya, during the periods of water shortage involve the fire and air by way of *Vāruṇi ahutis* to do *Vāyuviloḍana* or *Ākāśa-manthan* (atmospheric stirring) so that strong evaporation of water from ground or sea may easily take place to make the air reach the saturation point.[7] The clouds thus seeded artificially belong to the category of *Agnija* clouds. They produce lightning and thunder. This is why, the stimulation of rain through artificially seeded clouds invariably, as our experience had it, is followed by the lightning/sparks and thunder. It has also been noticed that this type of clouds takes one to fifteen days to precipitate. Therefore, it takes five to fifteen days to induce rain artificially through *Yajña* under tropical and continental conditions when the continental air is under turbulent sway.[8]

Soma (Moon)

In addition to the sun, the moon was also discovered as one of the factors that back-up the process of cloud-seeding. According to one of the Vedic seers, the vapours are held in the air under the impression of sun or moon.[9]

Such clouds as originated under the impression of the moon, or with the moon in their background were qualified by the Vedic sages as *somapṛṣṭhā* (i.e the clouds that are formed with the moon in their background.)[10]

Later meteorologists named these clouds as *pakṣaja,* i.e.

the clouds formed in different phases of the moon.

Varāhamihira, an ancient astrologer of this country has given a detailed break-up of the seeding and precipitating behaviour of these clouds.

Seeding of Pakṣaja clouds

The seeding of this type of clouds generally takes place during the period beginning from the bright half of *Mārgaśīrṣa,* through *Pauṣa, Māgha, Phālguna, Caitra* and *Vaiśākhā* when the moon enters the constellation of *Purvāṣ āḍhā, Uttarāṣāḍhā, Pūrvabhādraapada, Uttarabhādrapada* and *Rohiṇī,* etc. Thus the moon's presence in various asterisms in different months, as stated above, leads to the formation of various clouds.

Symptoms of cloud-seeding

The ancient Indian meteorologists carefully and minutely studied and investigated all such general and specific symptoms as indicated the seeding behaviour of clouds in the particular months marked by the presence of the moon in certain asterisms.

General Symptoms

The following have been determined as the general symptoms for cloud-seeding.

1. A delightful and cool breeze blowing from north, north-east or east.[11]
2. Clear sky.[11]
3. The sun and moon encircled by a glossy, bright and thick halo.[11]
4. The existence of bulky, large, glossy or smooth needle-like or razor-like embryos (*abhras*) having red hue or hues like eggs of crows (i.e. bluish) or of the peacock's neck in the sky.[12]
5. The pure and clear appearance of moon and stars in the sky.[12]
6. The auspicious dawn and twilight accompanied by a

rainbow, feeble thunder sound and the mock sun.[13]

7. The herds of animals and birds producing calm-voices in north, north-east or east.[13]
8. The planets being large, having glossy rays and uneclipsed moving to the north of asterism.[14]
9. Trees with the intact sprouts; men and animals happy.[14]

The Above mentioned signs are in general favourable for cloud-seeding.

Specific Symptoms

In addition to the above listed general symptoms, some specific periodic symptoms were also investigated which could foretell about the starting of cloud-seeding process. They are as under:

1. A reddening of the horizon at the time of dawn and twilight, the occurrence of clouds, halos around the moon and the sun and moderate snowfall point out the commencement of cloud-seeding process in the month of *Pauṣa.*[15]
2. In the month *Māragaśīrṣa,* all the symptoms remain the same as in the month of *Pauṣa* the only difference being that snowfall is replaced by cold.[15]
3. In the month of *Māgha,* the symptoms foretelling the cloud-formation are (1) a strong wind (2) the sun and moon dim by mist (3) bitter cold (4) appearance of clouds at the time of setting and rising of sun.[16]
4. In the month of *Phālguna,* a rough and violent storm, floating glossy clouds, an incomplete halo around the sun or the moon and the red or russet sun are found to be favourable conditions for cloud-seeding.[17]
5. In the month of *Caitra,* the persistence of winds, clouds, rain and halos favour the conception of a foetus by clouds (cloud-seeding).[18]
6. In *Vaiśākha* also the conception or cloud-seeding is promoted by the phenomena of winds, clouds, rain, lightning and thunder.[19]

The persistence of all these symptoms in the concerned time-periods shows that the rain-embryos (*abhras*) have been retained to be delivered in due course of time or say the clouds have been seeded to precipitate in due course of time.

The ancient meteorologists have also made in-depth studies to determine the actual period of precipitation/delivery of all such clouds as having conceived their respective features (rain-embryos) or have seeded in different seeding/conceiving periods due to moon's stay in various constellations.

The Period of Precipitation

The cloud that seeds under the moon's conjunction with particular constellation will precipitate after 195 days when the moon re-enters the same constellation.[20]

For instance, if the cloud begins to seed in the dark fortnight of *Pauṣa,* it will precipitate in the bright fortnight of *Śrāvaṇa.*[21] If the cloud begins to seed in the bright half of *Māgha,* the precipitation will take place in the dark half of *Śrāvaṇa.*[22] If, however, the seeding starts in the dark fortnight of *Māgha,* the precipitation will occur in the bright fortnight of *Bhādrapada.*[23] Similarly, if the seeding offsets in the bright fortnight of *Phālguna,* the precipitation will occur in the dark fortnight of *Bhādrapada.*[24] The clouds those begin to seed in the dark fortnight of *Phālguna* will precipitate in the bright fortnight of *Āśvin.*[25] Likewise, the clouds those conceive their foetuses in the bright and dark fortnight of *Caitra* will precipitate in the bright fortnight of *Āśvin* and the bright fortnight of *Kārtika* respectively.[26]

It is peculiar about the *Pakṣaja* clouds that should they start seeding in the dark fortnight, they will precipitate in the bright fortnight and vice-versa and should they start seeding in the day-time, they will precipitate at night and vice-versa, that which at dawn, in twilight and vice-versa, that which in the forenoon in late night and vice-versa, and that which in midnight in the mid-day and vice-versa.[27]

One more eccentric feature about the precipitation behaviour of these clouds is that they precipitate in the direction opposite to that of seeding. For instance, if they seed in the east, they will precipitate in the west and vice-versa. The same rule holds goods in case of other pairs of directions and also in case of the position of winds at two periods.[28]

Planetary Factors Leading to Cloud-formation

Apart from the sun and the moon which are the extra-planetary factors that control/determine cloud-formation on the earth, there are also some planetary factors that determine the distribution of precipitation and cloud behaviour on the earth. Most of the planetary factors leading to cloud-formation are geocentric. They include geographic conditions and geostrophic flow (earth turning motion). Owing to these geocentric factors, clouds are formed automatically by the sea-bearing winds (like a monsoon in India) in particular periods (called seasons) around the globe generating varying amounts of rain. The Vedic meteorologists have noticed the formation of clouds from seasonal sea-bearing winds as:

> 'Let the sea-bearing winds blow from all sides.'[29]

At another place, he observes as follows:

> 'Let all the directions fill with the sea-bearing winds which make the rain-embryos (*abhras*) float all around. Let the directions overcast with rain-embryos and the sea-bearing winds irrigate the earth with waters.'[30]

Such rain-generating clouds as are formed in various rainy seasons due to some geocentric reasons without any help from extra-planetary factors are called by the ancient meteorologists as *Brahmaja* i.e. created automatically in particular time periods. These clouds are generally devoid of thunder and discharge rain covering a vast region in that particular period.

De-clouding or De-seeding of Clouds

As is gathered from the name, de-clouding or de-seeding of clouds is the anti-phenomenon to that of cloud-seeding. De-clouding or de-seeding of clouds also takes place like that of cloud-formation. The process of de-clouding may be compared with the biological process of a miscarriage of foetus. Just as the miscarriage takes place in the living-beings, a cloud may also miscarry the foetus due to some reasons or others. The Vedic meteorologists were also familiar with the artificial means of de-clouding since they applied them to ward off rain. The subject rain-maker has also been de-seeding the clouds from time to time in order to ward off rain. He has so far conducted over 50 successful anti-rain experiments in different parts of India (the special one being done in Tripura for continuously one month) by way of artificially de-seeding the clouds.[31]

Symptoms of non-formation or non-seeding of clouds

The Vedic meteorologists investigated several natural symptoms, like that of cloud-formation, indicating the impairment of the process of seeding of clouds, or the miscarriage of foetus in clouds unless otherwise affected.

Some of the symptoms leading to de-seeding of clouds may be listed as under:

1. Fall of meteors.
2. The occurrence of a thunderbolt.
3. The occurrence of rain without clouds.
4. Governance of *Ketu.*
5. Dust-storm, the glow of horizon and tremor.
6. The appearance of the aerial city.
7. Planetary fight.
8. An unnatural phenomenon in rain such as the appearance of blood, etc.
9. A cross-bar of clouds at sunset or sunrise, i.e. a line of clouds standing across the sun at its rising or setting (*prighā*).
10. A glance of *Rāhu,* i.e. lunar or solar eclipse.

11. And unseasonal growth of flower all this indicates the impairment of seeding.[32]

Apart from the symptoms quoted above, the presence of excessive rain during the period of conception also bars the cloud from seed and the occurrence of rain exceeding the amount of 1/8 of a *droṇa* causes miscarriage of foetus.[33]

The clouds will also be destroyed in case a comet touches the Great bear, *Abhijit* asterism, the Pole star and the *Jyeṣṭha nakṣatra.*[34]

Sign of Rain-generating Clouds

Clouds undergo a change of colour and shape, while on the verge of yielding precipitation. For instance, clouds that resemble pearls and silver, *tamāla,* blue lily and collyrium in colour and are shaped like aquatic animals yield abundant water.[35]

Those clouds which are scorched by the fierce rays of the sun and fanned by a gentle breeze will yield torrential rain at the time of precipitation.[36]

Notes and References

1. See author (1993: art. 1.2.1.)
2. *ādaha svadhāmanu punargarbhatvamerire.*
   dadhānā nāma yajñiyam.
3. *agnir vā ito vṛṣṭim udīrayati. Kāt. S.* 2.11, Also *RV.* 1.164.47.
4. *prajāpati salilādā samudrād āpa iryannudadhimardayāti.*
   ūdīryata marutaḥ samudrataḥ tveṣo arko nabha ut pātyātha.
   AV. 4.12.2.

   Cf. also the following verses:

   kṛṣṇaṁ nīyānaṁ harayaḥ suparṇāḥ āpo vasānā divam utpatanti. asitavarnāḥ harayaḥ suparṇāḥ miho vasanā divam utpatanti. *TS.* 3.1.1.1
5. *rasāna raśmibhirādāya vāyunāyaṁ gataḥ saha*
   varṣatyeṣa ca yalloke tenendra iti smṛtaḥ. *BD.* 1.68.
   AV. 3.13.2.

6. See author (1993: 73-75)
7. *Ibid.*
8. *āpo bhadrā ghṛtamāpa āsannagniṣomau vibhṛtyāpa it tāh.* *AV.* 3.13.5
9. *ye parvatāḥ somapṛṣṭhā āpa uttānaśivarā.* *AV.* 3.21.10
10. *hlādimṛdudakśiva śakadigbhavo māruto viyadvimlam snigdhakṣita bahula priveṣa parivṛtau himmayukhārkau.*

    BS. 21.14
11. *pṛthubahulasnigdhaghanaṁ ghanasu cikṣ urakalohitābhrayutam.kākāṅḍamecakābhaṁ viyadviśuddhendu nakṣatram.* *BS.* 21.15
12. *surcāpa mandra garjita vidyut pratisūryakā śubhā sandhyā śaśi śiva śakrāśāsthāḥ śāntaravāḥ pakṣimṛgasaṁghāḥ.* *BS.* 21.16
13. *vipulāḥ pradakṣiṇacarāḥsnigdhamayukhā grahā nirupasargāh. tarvaśca nirupasṛṣṭāṇkurā naracatuṣpadā hṛṣṭāh* *BS.* 21.17
14. *pauṣe samārgaśīrṣe sandhyā rāgo'mbude sapriveṣāḥ nātyarthaṁ mṛgaśīrṣe śītaṁ pauṣe'tihimapātaḥ.* *BS.* 21.19.
15. *māghe prabalo vāyustuṣārakaluṣadyuti raviṣaṣāṇkau. atiśītaṁ saghanasya ca bhānorastodayau dhanyau. BS. 21.20.* Cf. also Kāśyapa on *BS.* 21.20.
    śitamabhraṁ tathā vāyu ścandrārkapariveṣaaṇam
    māghe māsi prīkṣeta ṣrāvaṇe vṛṣṭimādiśet.
16. *phalgunamāserukṣaścaṇḍaḥpavano'bhrasamplavāḥsnigdhāḥ pariveṣāścāsakalāḥ kapilastāmro raviśca subhaḥ*[23]

    Cf. also kāśyapa quoted by bhaṭṭotpala on *BS.* 21.21

    phalgune cātra saṁghātaṁ vṛṣṭistanitameva ca
    purovātāśca ye proktā māsi bhādrapade śubham.
17. *pavanaghana vṛṣṭiyukāścaitre garbhā śubhāḥ sapariveṣāḥ.* *BS.* 21.22

    Cf. also Kāśyapa quoted by Bhaṭṭotpala on *BS.* 21.22.

    bahupuṣpa phalā vṛkṣāvāḥ śarkarvarṣiṇah
    śitavarṣaṁ tathābhrāṇi caitreṇāśvayujaṁ vadeta.
18. *ghanapavana salila vidyutstanitaiśca hitāya vaiśākhe.* *BS.*21.22.

See also kāśyapa on *BS.* 21.22.

vahanti mṛdavo vātāḥ puraḥ śīghaṁpradakṣiṇāḥ
vaiśākhe tāni rūpāṇi kārttike māsi vartate.

19. *yannakṣatramupagate garbhaścandre bhavet sa candravaśāt pañcanavate dinaśate tatraiva prasavamāyāti* *BS.* 21.7.

20. *pauṣasya kṛṣṇapakṣeṇa nirdiśecchrāvaṇasya sitam.* *BS.* 21.9.

21. *māghasitotithā garbhāḥ srāvaṇa kṛṣne prasūtimāyānti.* *BS.* 21.10.

22. *māghasya kṛṣṇapakṣeṇa nirdiśedabhādrapadaśuklam.* *BS.* 21.10.

23. *phālgunaśuklasamuthā bhādrapadasyāsite vinirdeśyāḥ.* *BS.* 21.11.

24. *tasyaiva kṛṣṇapakṣodbhavāstu ye te'śvayukśukle.* *BS.* 21.11.

25. *caitrasitapakṣjātāḥ kṛṣṇe'svayujasya vāridā garbhāḥ caitrāsita sambhūtāḥ kārtika śukle'bhivarṣanti.* *BS.* 21.12.

26. *sitapakṣabhavāḥ kṛṣṇe śukle kṛṣnā dyusambhavā rātrau naktaṁ prabhavāścāhani sandhyājātāśca sandhyāyām.* *BS.* 21.8.

27. *purvodbhūtāḥ paścādaprotthāḥ prāgbhavanti jīmūtāḥ śeṣāsvapi dikṣvevaṁ viparyayoḥ bhavati vāyośca.* *BS.* 21.13.

28. *saṁ no iṣiro vātu vātaḥ.* *RV.* 1.35.4.

29. *samutpatantu pradiśo nabhasvatiḥ samabhrāṇi vātajutāni mahaṛṣbhasya nadato nabhasvāto vāśrā āpaḥ pṛthiviṁ tarpayantu.* *AV.* 4.12.1.

30. See for detail author (1993:79)

31. *garbhopaghātaliṅgānyulkāśnipāṁsupātadigdāhāḥ kṣitikampa khapura kīlakaketu graha yuddha nirghātāḥ. rudhirādvṛṣṭi vaikṛtaprighendra dhanuṁṣi darṣanaṁ rāho ityutpātairetaistri vidhaiścānyairhato gorbhaḥ.* *BS.* 21.25.26.

Cf. also Garga quoted by Bhaṭṭotpala on *BS.* 21.26.

aśmavarṣaṁ tamo varṣam māṁsaśoṇita varṣaṇam
ulkānirghāt kampaśca vajrapātastathaiva ca

pariveṛaḥ paridhiyo vāsavasya dhanuṁṣi ca
anabhrastanitaṁ varsaṁ diśām dāhastathaiva ca.

anārtavaṁ puṣpaphalaṁ vārṇiyeṣu varṣaṇam
graheyuddheṣu ghoreṣu hatān garbhān vinirdiśet.

32. *garbhasamaye'ti vṛṣṭirgarbhā bhā vāya nirnimittakṛtā*
droṇaṣṭāṁśe'bhyadhike vṛṣṭe garbhaḥ śruto bhavati.BS. 21.34.

33. *munīnabhijitaṁ dhruvaṁ maghavataśca bhaṁ*
saṁspṛśan śikhi ghanavināśakṛt. *BS.* 47.12.

34. *muktārajatnikāśāstamāla nilotpalañjanābhāsaḥ*
jala cara sattvā kārā garbheṣu ghanāḥ prabhūtajalāḥ.
BS. 21.23.

35. *tivra divākara kiraṇā bhitāpitā mandamārutā jaladāḥ*
ruṣita iva dhārābhirvisṛjantyambhaḥ prasavakāle. *BS.* 21.24.

4

Vidyut (Electrical Charge)

The second phase of change that the *āpaḥ* (moisture) undergo is the production of electrical charge (*Vidyut*) in the clouds. The mechanism by which a cloud becomes electrically charged is though not perfectly known to the modern meteorologists, yet their speculations hold that the occurrence of very violent motions within a large cloud may lead to the production of electrical charge.[2] Since, according to them, when two substances are in a violent motion with respect to each other, the parts are often found to be electrically charged. The often-quoted example to support this postulation is the production of electrical charges on an ebonite rod and a flannel by vigorous friction. Although both were uncharged (or electrically neutral) originally, but after the violent action both the rod and flannel are found to be possessed of electrical charges.[3] Though the explanation is given by modern scientists based on the speculation of violent motion in the clouds in respect of occurrence of electrical charge in the clouds seems to be plausible and reasonable to the scientific eyes, it is still shrouded in uncertainty.

On the other hand, the Vedic meteorologists had a clear cut vision. According to them, all sorts of clouds cannot possess electrical charge, but only the clouds formed from vapours evaporated by the radiation heating from the sun become electrically charged. They have given a very beautiful reason behind this phenomenon. Accordingly, the waters on being evaporized by radiation heating from the sun get electrically charged. So, they are called *āpaḥ*).

A verse of an *Ātharvaṇa* seer may in this regard be read as under.

> **'Because you are sent forth (evaporized) by the *varuṇa* (radiation heating from the sun) and you have condensed so well, *Indra* (electrical charge) possesses you. You are, therefore, called *āpaḥ* (charged clouds).[5]**

These *āpaḥ,* when set to move violently, are called *pravatas* (or floating clouds). Since the electrical charge is produced in the floating clouds (*pravatas*), it is called by Vedic seers as *pravatonapat,* i.e. the daughter (*napāt* of *pravatas* (of floating clouds). For instance:

> 'I salute you O' electrical charge. You take birth in the floating clouds.'[6]

The electrical charges positive as well as negative act as a nucleus for foregathering of clouds. The *Ṛgvedic* seer observes this fact as follows.

> 'O' the daughter of floating clouds, i.e. the electrical charge! let me salute you and your forms-*hetu/agneya,* or negative as well as *vapu/somīya* or positive. We know your residence which is *āpaḥ.* Hiding therein, you act as nucleus for fore-gathering the clouds.'[7]

Indra

The Vedic seers called the electrical charge produced in the clouds formed from vapours evaporated by radiation heating from the sun as *Indra.* Since the origin of *Indra* takes place in the violently moving *āpaḥ* transferred by the radiation heating from the sun, the *Indra* came to be known as the representative of the sun, the celestial fire in the mid-space. Actually, the electrical charges in the clouds lead them to yield rain. This is why, *Indra* was called as the rain-god, or the physical force operating in the mid-sphere for the promotion of rain.[8] He is often remembered by the Vedic sages as the killer of *Vṛtra* (cloud). This points out that the production of discharge causes the rainfall. In his exploits, as gathered from the vision of Vedic seers, the *Indra* is often assisted by *Marutas.* When *Indra* kills *Vṛtra,*

Marutas sport around him.[9]

In fact, *marutas,* or air particles give rise to ionization that leads to the discharge/spark between the oppositely charged parts of a cloud or two clouds or charged cloud and neutral earth. This is why *marutas* (-vely) are eulogized by the Vedic seers as the conduction maker of the charge.

Notes and References

1. *Aeolus* (1952:84)
2. *Ibid.* (71)
3. *Ibid.* (83)
4. See fn. I *supra*
5. *yat preṣitā varuṇenācchibhaṁ samvalgata tadāpnodindra vo yatistasmādāpo anuṣṭhan.* *AV.*.13.2.
6. *namastepravatonapād yatastāpaḥ samuhasi.* *VS.* 36.21.
7. *pravatonapāt nāma evāstu tubyaṁ namaste hetaye vapuṣe ca kṛṇmaḥ vidma te dhāma paraṁ guhā yat samudre antarnihitāsi nābhiḥ.* *AV.* 3.13.3.
8. See *BD.* 1.68.
9. *maruto ha vai krīḍino vṛtraṁ haniṣyantam indraṁ āgatam abhitaḥ paricikruḍuḥ.* *Ś.Br.* 2.5.3.20.
10. *vidyunmahasaḥ.* *RV.* 5.54.3.

5

Stanayitnu (Thunderstorm) or Lightning and Thunder

The third change of phase of *āpaḥ* takes place in the form of *stanayitnu,* i.e. thunderstorm, or lightning and thunder.[1] In this phase *vidyuta* (electrical charge) in the clouds turns into light and sound. The factor that works behind the conversion of an electrical charge into thunderbolt in the cloud is called by the Vedic meteorologists as *apsarā.* Its meaning may be made more clear with the help of its etymology given by the ancient Vedic scholar Yāska. He registers the concept behind the origin of this word as *apsarā, apsāriṇī, apsu, sāriṇī*[2] i.e. *apsarā* is that which gets induced in the *āpaḥ* i.e. electrically charged vapours or the vapours charged by negative and that which neutralizes the charges. This way, *apsarā* is the positive charge induced by the negative charge. On the other hand, *Indra* is the negative charge. It may be pointed out here that whenever the author uses the term electrical charge in this book, he means it only by the negative charges. Thus the contact between the opposite charge, i.e. *Indra* and *Vṛtra* creates a short circuit between their conductors and unleashes a high amplitude current lightning followed by acoustic shock waves caused by the sudden rise in the air temperature along the lightning's path. This phenomenon of lightning and thunder was called by the Vedic meteorologists as the *Vajra* of *Indra or aśani.*

The *Ṛgvedic* seer observes this phenomenon as under:

'Lightning is created between the two charges.'[3]

According to the *Taittirīya* seer:

'When it (*agni*) is expanded into two forms negative and positive, its nucleus is shattered or say it is neutralized and so thunderbolt comes into being.'[4]

Kapiṣṭhal seer also holds the similar view. In his opinion:

'The ionization of air particles help produce the discharge between the opposite charges and so nucleus of *agni* is shattered. This discharge causes the origin of thunderbolt.'[5]

The acoustic shock waves (thunder sound) that follow the lightning was also named by the Vedic seers as *Rudra.* Since it is also created along with the lightning, it is also described as the form of *agni, rudro' agniḥ*[6]

Notes and References

1. Vedic meteorologists take note of this change, hence they welcome the change in the following words:
   namaste astu vidyute namaste stanayitnave.
2. *Nir.* 5.14.
3. *aśmanor madhye agniṁ jajāna.*
4. *tasya tantasya hṛdayaṁ ācchindam so aśnir abhavat.*
   *TB.*1.1.3.12.
5. *tasya marutaḥ stanayitnunā hṛdayam ācchindana sa divyāśnir abhavat.* (6.7)
6. *Kāt. S.* 8.8; 24.6; *KS* 42.6

6

Precipitation

Āpaḥ cycle is completed with their last change of phase as precipitation, in which they (water) fall back to the earth in the forms of *kāra* (hailstones), *dhāra* (rain-drops) and *tuṣ āra* (snow-flakes).

***Dhāra* (*Rain-drops*)**

The *dhāra* form of precipitation includes in it also the *śikara* type of rain, i.e. the precipitation of drizzle drops.

The precipitation in the form of *dhāra* was studied variously by the Vedic meteorologists in the context of various types of clouds. For instance, in the context of *agnija* (fire) clouds, it was seen as a battle between *Indra* and *Vṛtra,* popularly known in the Vedic history as '*Indra Vṛtrāsura Yuddha'.* In the context of *Pakṣaja* and *Brahmaja* clouds, it was seen as a time-bound process.

***Indra Vṛtrāsura Yuddha* (Battle between *Indra* and *Vṛtra*)**

Indra Vṛtra Yuddha is a figurative description of the process of precipitation of rain-drops from *agnija* clouds. *Indra,* as stated above, is a negative electrical charge occurring in the clouds. *Apsarā* is an opposite charge i.e. positive charge that rushes towards *Indra* (negative charge). In fact, where there is *Indra Loka,* there enters *Apasarās.* This is why, in the symbolic description of the Vedas and the later *Pauraṇic* mythological tracts, *Apsarās* have been shown as residing with *Indra.* Contact between the opposing charges, i.e. *Indra* and *Apsarā* creates a short circuit and unleashes a high amplitude current, lightning. This bolt of current was called the *vajra* of *Indra*. This *vajra* pierces the cloud into many pieces and makes it ultimately fall on the earth. The *Vṛtra's* falling on the earth into pieces is the falling of rain-drops. This is how the process of *dhāra* precipitation or rain-fall from the *agnija* clouds is accomplished.

Dhāra precipitation from Pakṣaja clouds

Precipitation from *Pakṣaja* clouds i.e. the clouds formed in a particular period under the moon's conjunction with some specific constellation is a time-bound process. Generally, the clouds that seed under the moon's impression become ready to precipitate after a period of six and half months, i.e. after 195 days. The whole process has already been discussed above in detail under the head *'The period of precipitation'* in chapter-4.

The Estimation of Dhāra Precipitation

The amount of precipitation from *Pakṣaja* clouds is subject to the presence of five factors, *viz.* the wind, water, lightning, thunder and cloud and sometimes depend upon the constellations and planets under the impression of which the cloud is seeded.

For instance, if the star at which the cloud-seeding is taking place to be aspected by the benefic planets in conjunction with the sun or the moon, there would be copious rain.[1]

If the cloud under the process of seeding pours down excessive rain even at the seeding period, it will produce drizzle at the time of precipitation.[2]

To be more exact regarding the *dhāra* precipitation, it is maintained that if the cloud-seeding is generated in the presence of five factors, *viz.* wind, water, lightning, thunder and cloud, the amount of rain-fall during precipitation period will be around a *droṇa*. If in the presence of the wind alone, three *āḍhakas;* in the presence of lightning alone, six *āḍhakas;* in the presence of clouds, nine *āḍhakas* and in the presence of thunder alone, twelve *āḍhakas.*[3]

But sometimes, we face differing views regarding the amount of rain-fall likely to occur from the clouds seeded in the presence of various factors. For instance, according to Vṛddha Garga quoted by Bhaṭṭotpala on *BS* 21.32 if the conception takes place in the presence of wind, it will rain measuring the amount of one *āḍhaka* while rest of the measurement remains the same.[4]

The occurrence of rain-fall in various asterisms unless the said asterisms are otherwise afflicted also go a long way to help determine the amount of rain-fall in the precipitation period. For instance, if it rains in any one of the asterisms, *viz. Hasta, Pūrvāṣāḍhā, Mṛgaśiras, Citrā, Revati* and *Dhaniṣṭhā*, the quantity of rain-fall in the season will be 16 *droṇas*; in *Śatabhiṣaj*, *Jyeṣṭhā* and *Svāti,* it will be 4 *droṇas*; in *Kṛttika 10 droṇas;* in *Śrāvaṇa*, *Maghā*, *Anurādhā*, *Bharaṇi* and *Mūla* 14 *droṇas*, in *Pūrvaphalgunī* 25 *droṇas*; *Punarvasu, Viṣākhā* and *Uttarāṣāḍhā* 20 *droṇas*; in *Āśleṣā* 13 *droṇas*; in *Uttarabhādrpada, Uttarāphalgunī* and *Rohiṇī* 25 *droṇas*; and *Ārdrā* 18 *droṇas.*[5]

It may be noted here, should the said constellation be afflicted by the Sun, Saturn, *Ketu* or Mars or threefold portents, there would be no rain in the season. On the other hand, should the asterism be aspected by benefic, i.e. Mercury, Jupiter and Venus, the precipitation will surely take place.[6]

One more view to get the estimate of the amount of rain-fall in the rainy season was also prevalent among the meteorologists. Accordingly, the amount of rain-fall that was likely to occur in the entire rainy season would be estimated by the amount of rain-fall recorded in the beginning of the rainy season headed by *Purvāṣāḍhā* after the full moon in the month of *Jyeṣṭha.*[7]

NB: According to *Parāśara*, a *droṇa* is four times the *āḍhaka* and *āḍhaka* is the capacity of a circular vessel whose diameter is 20 *aṅgulas* and depth 8 *aṅgulas.*

The Range of Dhāra Precipitation

The investigations of Vedic visionaries also led them to determine the range of the area to be covered by rain pouring clouds formed to the accompaniment of certain meteorological factors named the wind, water, lightning, thunder and cloud. For instance, according to the investigations of the ancient meteorologists, if the cloud begins to form to the accompaniment of all the five meteorological factors mentioned above, it will pour rain over an area of 100 *yojanas,* if the cloud-seeding takes place to the accompaniment of four, it will rain over an area of 50 *yojanas*; if to the accompaniment of three, 25 *yojanas*; if to the accompaniment two, 12 ½ *yojanas;* and if to the accompaniment of only one, 6 ½ *yojanas* around.[8]

NB: Though the above-cited rules of the estimate of amount and range of *dhāra* precipitation have been given by the Vedic meteorologists in the context of *Pakṣaja* clouds, the same rules hold good in the context of *Agnija* and *Brahmaja* clouds.

Dhāra from Brahmaja clouds

The *dhāra* precipitation from *Brahmaja* clouds is also, as stated earlier, a time-bound process. The clouds seeding in a particular period through sea-bearing winds over different parts of the globe will precipitate in the normal course of events, i.e. unless otherwise afflicted through artificial means in the fixed time-span over the fixed range of area. Most of the studies conducted in modern times to determine the distribution of rain-fall over the globe are based on the behaviour of *dhāra* precipitation from the *Brahmaja* clouds.

Measuring devices for Dhāra precipitation

The measurement of *dhāra* precipitation appears to have prevalent in India since the very inception.

Varāhāmihira quotes the Vedic sage *Parāśara* in

connection with his measurement of one *āḍhaka* of rain-fall. Accordingly, *Paraśara* measured one *āḍhaka* of rain-fall through the circular vessel which would have a diameter of 20 *aṅgulas* and depth of 8 *aṅgulas.*[9]

Later on, another gauge for measuring rain-fall was developed with slight parametrical variations. For example, Varāhāmihira refers to a gauge which has a diameter of one cubit. Its fifty *pala* time capacity was fixed as a unit for storing one *āḍhaka* of rain-water.[10]

Furthermore, four-time of *āḍhaka* was measured as a *droṇa.*[10]

The earliest references to rain gauging are found in Pāṇini's *Aṣṭādhyāyī,* such as *varṣa-pramāṇe.*[11] *goṣpada* is referred to as the smallest measure of rain-fall.[12]

The references to rain-gauging and rain-gauges mark the significant advancement of meteorological sciences in ancient times.

Kāra (Hailstones)

The second type of precipitation studied by Vedic meteorologists was in the form of *Kāra.*

Formation of Karakas (Hailstones)

With regard to the formation of *Karaka* (hailstone) precipitation from *agnija* clouds, we can show the vision of an *Ayurvedic* seer. According to him, *divya vāyu* (mid-spatial air), i.e. the violent motion in the clouds and the *divya agni* (mid-spatial fire) i.e. the electrical charging of the clouds provide the setting for the hailstones.[14]

The *kāra* precipitation is also said to occur from the *pakṣaja* clouds. according to a study made with regard to the precipitation behaviour of *Pakṣaja* clouds it is not necessary that a super-cooled *pakṣaja* cloud will precipitate in due course of time, i.e. its precipitation period. Sometimes, owing to such reasons as arise from the afflictions of the constellations or others, a super-cooled cloud is not able to deliver its lot at the time of its delivery and so the delivery or precipitation is postponed or put-off for an indefinite period. Due to this postponement of delivery, the water droplets contained in a cloud grow thick and hard, just like the milk of milch cattle grow thick and hard if kept for too long.[15] This is way, the precipitation of hailstones generally takes place in odd periods either during its onset or the closing of the precipitation period or sometimes during off-seasons or uncommon times of the year.

The whole phenomenon of precipitation of hailstones may be described in more modern scientific terms as:

The cloud that which fails to precipitate in due course of time due to some reason or others are forced to move higher up in order to give way to others that are to deliver their foetuses in their respective turns. The more the cloud is updrafted, the more it finds the opportunity to reach the freezing point. This upliftment of the cloud ultimately gives rise to the phenomenon of hailing.

Sometimes, the aspect of malefic planets is also regarded to cause hailing. For instance, if the constellation at which the cloud is seeded be aspected by the malefic planets, there would be a precipitation of hailstones along with thunderbolt from the (*Pakṣaja*) clouds.[16]

It may be noted here that the precipitation of hailstones never takes place from *Brahmaja* clouds. Only the *agnija* and *Pakṣaja* clouds yield hails that too often in uncommon times of the year.

Tuṣāra (Snow-flakes)

Tuṣārapāta or *himphāla,* i.e. precipitation of snow-flakes generally takes place on high altitudes. This is why, the Vedic sages named high altitudes experiencing the precipitation of snow-flakes as *Himavān* or *Himālaya*, etc. etc.

Notes and References

1. *śaśini ravau vā śubha saṁyutekṣite bhūrivṛṣṭikarāḥ. BS.21.33.*
2. *Visṛjati yadi toyaṁ garbhakāla'tibhūri prasavasamayamitvā śikarāmbhaḥ karoti.* *BS.* 21.37.
3. *droṇah pañcanimittaḥ garbhe triṇyāḍhakāni parena taḍ vidyutā nāvābhraiḥ stanitena dvādaśa prasave. BS.* 21.32.
4. *vāte tu āḍhakaṁ vindyāt stanite dvādaśāḍhakam navāḍhakaṁ talhā'bhreṣu dyoteṣu ṣaḍāḍhakam nimitta pañcakopete droṇaṁ varṣati vāsavaḥ.*

   BS. 32.1.

   hastāpyasaumya citrā pauṣṇa dhaniṣṭhāsu droṇāḥ śatabhiṣ againdrasvātiṣu catvāraḥ kṛttikāṣu daśa.sravaṇe maghānurādhābharaṇī mūleṣu daśa caturyuktāḥ phalgunyāṁ pañcakṛtiḥ punarvasau viṁśatidronāḥ aindrāgnyākhye vaiśve ca viṁśati sarpabhe daśatrayadhikāḥ āhirbudhnyāryamṇa prājāpatyeṣu pañcakṛtiḥ.

   pañcadaśāje puṣye ca kīrtitā vājibhe daśa dvau ca raudreṣṭ ādaśa kathitā droṇa nirupadraveṣvete. *BS.* 23.6.9.

   Cf. also verses of *Samāsasaṁhitā* quoted by Bhaṭṭotpala on *BS.* 23.6.9.

   The verses go like this:

   daśayuktā dvikṛtakhatithirasāṣṭadigviṣ ayarāmajalatithibhiḥ tithirasarasaiśca virasāḥ sadaśakṛtāḥ ṣaḍvihināśca jalaṣaṭ kadaśakasahitā jalarasayuktāḥ ṣaḍūnāśca viṣayatithiṣaṭka sahitāśvinyādiṣu jaladroṇāḥ.

5. *ravi ravisuta ketu piḍite bhe kṣititanayatri bhidhādbhutā hate ca bhavati ca na śivaṁ na cāpi vṛṣṭiḥ śubhasahite nirupadrave śivaṁ ca.* *BS.*23.10.

   jyaiṣṭhyāṁ samatītāyāṁpūrvāṣāḍhādapi sampravṛṣṭena śubham aśubhaṁ vā vācyaṁ pariṇāmaṁ cāmbhasastjjñaiḥ. BS. 23.12.
6. *pañca nimittaiḥ śatayojanaṁ ladaradhārdhamekā hānyātaḥ varṣ ati pañcanimittādrūpeṇaikena yo garbhāḥ.* *BS.* 23.14.
7. *same uiṁśāṅgulenāhe dvicatuṣkāṅgulocchrite bhāṇḍe varṣati*

sampūrṇaṁ jñeyamāḍhaka varṣaṇam. *BS.* 21.32.

8. *hastaviśālaṁkundakamadhikṛtyāmbupramāṇa-nirdeśaḥ pañcāśatpalam āḍhakamānena minuyājjalaṁ patitam.BS.* 23.2.

9. Cf. Bhaṭṭotpala on *BS.* 23.2 : *catubhirāḍhakairdroṇaḥ*

10. *Pān.* 3.4.32.

11. *Ibid.* 6.1.145.

12. *divya vāyva agnisaṁyogāt saṁhatāḥ khāt patanti yāḥ śilāprakārabaddhāstāḥ karakāamṛtopamāḥ.*

13. Cf. *garbhapuṣṭaḥ prasave grahopaghātādi bhiryadinavṛṣṭaḥ ātmīya garbha samaye karakāmiśraṁ dadātymbhaḥ.BS.* 21.35.

Also

kāṭhinyaṁ yāti yathā cirakāladhṛtaṁ payaḥ payascinyāḥ kālātītaṁ tadvatsalilaṁ kāṭhinyamupayāti. *BS.*21.36

krūrgrahasaṁyukte karkāśani matsya varṣadā garbhāḥ. *BS.* 21.33.

7

Avagraha (Anti-Rain)

The ancient meteorologists were as conscious with the concept of anti-rain (failure of rain) as with the precipitation of rain. The phenomenon of anti-rain was termed as *anāvṛṣṭ i, varṣapratibandha, varṣānighāta,* etc. etc. Pānini refers to it as *avagraha.*[1]

Periods of Avagraha

The ancient meteorologists sorted out the periods which usually witness no rain. For instance, four days beginning on the eighth lunar day of the brighter fortnight of the *Jyeṣ ha* month are considered to be a period of anti-rain, since that period is dominated by the air.[2]

Apart from this, the period for cloud-seeding beginning from *Jyeṣṭha* till *Kārtika* was considered as an uncommon time of the year (in Indian context) or the off-season for precipitation.[3]

Causes of Avagraha

The ancient meteorologists also sorted out some extra-planetary effects causing the phenomenon of anti-rain.

Accordingly, should an asterism beginning with the *Purvāṣādhā* after the full-moon in the month of *Jyeṣṭha* be aspected by the Sun, Saturn, *Ketu,* or Mars, or afflicted by three-fold portents (*trividha utpātas*)[4] there will be no rain in the entire season.[5]

Moreover, if in the brighter fortnight of the *Jyeṣṭha* month there be rain in four lunar mansions of *Svāti,*

Viśākhā, Anurādhā and *Jyeṣṭhā*, the months of *Srāvaṇa, Bhādrapada, Āśvin* and *Kārtika* respectively would witness no rain.[6]

Notes and References

1. *pān.* 3.3.51.

2. *jyeṣṭhasite'ṣṭmyādyāścatvāro vāyudhārṇā divasāḥ.*
   BS. 22.1.

3. Cf. Kāśyapa's view quoted by Bhaṭṭotpala on *BS.* 22.2.
   jyeṣṭhasya śuklāṣṭamyāṁ tu nakṣatre bhagadaivate
   catvāro dhārṇā proktā mṛduvātā samīrītāḥ.

4. Threefold portents have been enumerated as:
   i. Celestial ones like the eclipse of planets.
   ii. Mid-spherical ones like the glow of horizon, fall of meteors and wind-cyclones, etc.
   iii. Terrestrial ones arising from moving and stationary objects.
   divyaṁ graharakṣavaikṛtamulkānirghāta pavana pariveṣāḥ gandharvapura purandara
   cāpādiyadāntarikṣaṁ tat. *BS.* 46.4.
   bhaumaṁ sthiracarabhavam. *BS.* 46.5.

5. *ravi ravisuta ketu piḍite bhe kṣiti tanaya trividhādbhūtāhateca bhavati ca na śivaṁ na cāpi vṛṣṭiḥ śubha sahite nirupadrave śivaṁ ca.* *BS.* 23.10.

6. *tatraivā svatyādye vṛṣṭe bhacatuṣṭye kramānmāṣāḥ srāvaṇpūrvā jñeyā pariśrutā dhārṇāstāḥ syuḥ.* *BS.* 22.2.
   Cf.also Kāsyapa's views quoted by Bhaṭṭotpala on*BS.* 22.2.
   svātau tu śrāvaṇaṁ hanyādvṛṣṭe' dhendrāgnidaivate bhādrapade tvavṛṣṭiḥ syānmaitre cāśvayuje smṛte aindre te kārtike tvevaṁ vṛṣṭe vṛṣṭiṁ nihanti ca.

8

Vāyu (Air)

Air mass is the third dominant factor that accounts for the creation of meteorological phenomenon or the various weather conditions on the earth. The part played by air in creating the meteorological phenomena, or weather conditions on the earth are quite obvious from the foregoing discussions. Here one can easily make out that right from the process of cloud-formation down to precipitation, air plays a significant role in the materialisation of various phenomena. The air is, in fact, the *prāṇa,* as it is called by the Vedic seers, a life-giving/vital factor in the universe without which nothing could move in nature whether it is vaporization; or cloud- formation, or lightning, or thunder, or whatnot.

Since air comes into being first of all after *ākāśa,* it pervades fire and water that ensue it. This is why, neither the water alone, nor the fire alone can form weather on the earth, but it is the air all the way that makes the *āpaḥ* cycle move or that transfers the conditions of hotness or dryness to the various parts of the globe.

Variables of Air

Various types of winds, storms, or hurricanes may be studied as the variables of air.

Winds

The meteorologists of ancient days identified various types of winds blowing through various layers.

These winds are (1) *āvaha* (2) *pravaha* (3) *saṁvaha* (4) *udvaha* (5) *vivaha* (6) *parivaha* (7) *prāvaha.*

Āvaha is the lowest layer of the wind on the surface of the earth. The *Sūrya Siddhānta*[1] says that the *āvaha* or provector wind impels the planets towards their epics.

Pravaha is the layer above *āvaha*, which is said to have sustained the celestial river *Ganga* known as *Mandākinī.* Above this layer is the *Udvaha* which is the third one in the series of seven. In the fourth layer comes the wind is known as *parivaha* which is virtually in the middle of seven. In the fifth stage and above the layer of *parivaha* is the *prāvaha. Vivaha* is the sixth and the *saṁvaha* is the seventh layer of air. These layers roughly come within the area of six hundred miles above the surface of the earth.

Winds of Strong Intensity or Reśma

A wind of strong intensity often referred to by the term *reśma* in the Vedic *Saṁhitās.*[1] This term is similar to a hurricane or storm which is applied to denote the wind of great intensity in the modern days. The origin of these various types of winds has been assigned to the fire by the Vedic meteorologists.

Notes and References

1. *AV.* 6.102.2 & 25.2.1.

9

Agni (Fire)

Agni (fire) *Agni* (fire) forms the phenomenon of temperature in the atmosphere and also accounts for the tropical and continental conditions in the weather of the earth. It is the main cause of cyclones. It stirs the air to flow in the shape of winds and hurricanes, etc. Keeping in view the same fact, the Vedic seer observes that all sort of motions and air flows in the universe is caused by the fire.[1]

Physical *Agni* is described by the Vedic seers to have occupied the entire universe in its various forms. In its latent form called as fire, it is present in the terrestrial sphere. In its violent form called as *Indra* (electricity), it is present in the midsphere. In its ionized form called as the sun, it is present in the celestial sphere. Thus it affects the atmosphere of the earth variously through its various forms. For instance, the terrestrial fire is used to modify weather by sending forth the oblations offered to it into the midsphere to augment the power of deficient elements.

The mid-spatial *Agni* helps the clouds precipitate in various forms, *viz.* drizzle drops, rain, hailstones and snowflakes.

The celestial *Agni* leads to the evaporation process. The various roles of *Agni* in various forms have been detailed in the second part of this book. So, it won't be proper to repeat the same here.

Notes and References

1. *agninā vāyuḥ* (*dīpyate*) *Ś.Br.* 10.6.2.11.

10

Weather Control on the Earth

So far we have taken into account all those factors that account for the creation of meteorological phenomena and various weather conditions on the earth or say the formation of the atmosphere on the earth. These factors may be known as the formative ones. Apart from these formative factors, there are also some controlling factors that affect the change in the atmosphere of earth from outside. Those factors that affect the atmosphere of the earth are studied here in the name of extra-planetary control of the weather.

Extra-planetary Control of Weather

The Planetary weather, though formed of the factors discussed in detail above is controlled to some extent by the movement of stars. Planets, constellations and comets, etc. Since, the stars, planets, constellations and comets in a particular solar system owe their birth to one particular solar nebula (*hiraṇyagarbha*) and they are sustained by the gravitational pull[1] of one another. In other words, they gravitate towards each other. Thus by way of the gravitational pull, i.e. either by being gravitated to others or by gravitating others they affect at large each other in many aspects. But here our main concern is the study of their meteorological effect on our planet.

Ancient meteorologists made in-depth studies to evaluate the effect of the heavenly bodies in the creation of meteorological phenomena on the earth. To study the extra-planetary effects, the Vedic seers developed a new branch of science popularly known as the *Jyotiṣa* or Astronomy, as is called these days. By way of this branch of science, they studied the relationship between the members of solar-system and what happens on the earth. They calculated the movements of the sun, moon and planets as they revolve in their orbit from a geocentric point of view, that is as if viewed from the earth. The circle in which the earth moves was found attended by twenty-seven constellations. Of these, each successive group consisting of two and a quarter part of the third constellation represents the elements of fire, earth, air and water respectively. Thus the total groups formed are twelve, and so a fictitious demarcation of 12 zones, each one consisting of a group of two and a quarter part of the third one was made in order to be particular about the assessment of the effect of each zone dominated by one of the four elements.

Now we shall detail hereunder the meteorological effects of the heavenly bodies on the earth as studied by our ancient sages.

Sun (Ravi)

The sun is a star composed mainly of ionized hydrogen and helium. It has been identified as a high-temperature zone. Its temperature at photosphere is about 6000^0c, at chromo-sphere about 32400^0c, at transition region about 324000^0c and corona about $2,700,000^0$c hot enough to emit X-rays. It radiates intense light and heat on the earth. So, if the year, month or day is presided over by the sun, the weather of that particular year, month or day will remain warm (even in winter).[2]

Astronomically also the sun dominates the fire element.

Hence, the sun's ruling though causes a lot of vaporization and hotness in the atmosphere, yet the air remains unsaturated. This leads to scanty clouding and scarce rainfall.[3]

Moon (*Soma*)

The moon is the only satellite of the earth. The moon is deficient in *vasus,* i.e. it has no atmosphere as its gravitational power is too weak to hold down *vasus*[4]. The moon along with the sun is responsible for the tides, as the moon is held by the earth and the earth is held by the sun. But moon being nearer to earth than sun exerts a greater effect on the earth. Astronomically, the moon dominates the water element. This is why, under the rulership or dominance of the moon, the evaporation takes place enough to give rise to thundering clouds often yielding heavy rain.[5]

In addition, the moon's diverse movements have been observed to affect the weather diversely. For instance, if the moon moves to the north-east of *Rohini,* there will be good and timely rains.[6] When the moon touches *Rohiṇi* while moving to the north, there will also be copious rain giving rise to a lot of disasters. On the contrary, when it moves northward without touching *Rohiṇi,* there will be abundant rain, but no disaster whatsoever.[7]

The moon's being aspected by benefics, when it occupies the 7th house from Venus, or 5th, 7th or 9th from Saturn during the rainy season, adds to its power in the positive direction and hastens the materialization of the resultant meteorological phenomenon, i.e. rain.[8]

Though the moon, by nature, is a rain-promoting factor, but when it is eclipsed by the strong planets like Mercury or Saturn or the *Ketu* (a comet) that promote anti-rain phenomenon on the earth by their influence, the rainlessness, or famine/drought will take place due to the dominance of anti-rain causing factors. That is why, ancient meteorologists studied the moon's eclipse by Saturn, Mercury or *Ketu* as consequent upon famine/drought or rainlessness.[9]

Mars (*Maṅgala*)

Mars is a close neighbour of the earth. It is smaller in

size than the earth. It has a thin atmosphere. It is a barren place covered in the pink soil and boulders, as per modern scientific investigations. Long ago it was more active. Its surface is marked with dormant volcanoes. Its volcanic activity rendered it a dry land. Astronomically, it dominates the fire element. Hence its effect on the earth also causes the conditions of evaporation and hotness. This is why, the ancient meteorologists investigated that if the year is ruled by Mars, the clouds, due to heavy evaporation from oceans, though appear thick and piled up in the sky, don't yield sufficient rain.[10]

Its movement via different constellations generally accounts for the phenomenon of rainlessness. For instance, should the Mars pass through the southern side of *Rohiṇi,* the rain would be marred.[11] If it makes a move through *Rohiṇi, Srāvaṇa, Mūla, Uttaraphalguni, Uttarāṣāḍhā, Uttrabhādrapadā, Jyeṣṭhā* it will even deseed the clouds.[12] If it is retrogressed from the 10th, 11th or 12th asterism counted from the one wherein it rises, it will give effect to the conditions of rainlessness.[13] If Mars commences its retrograde motion from 17th or 18th constellations, it will give rise to the condition of drought.[14] Should Mars not commence its retrograde motion from the middle of the *Maghā* asterism, there would be no rain.[15] If it passes through *Viśākhā* having passed via *Maghā*, the famine will take place.

In spite of its anti-rain effect on the earth, it is also noticed that some of its retrograde motions heralded food or immediate rain. For instance, the retrogression of Mars from the 15th and 16th asterism counted from excluding the one wherein it rises after the eclipse, gives effect to immediate rain.[16] If Mars is retrogressed from the 13th or 14th asterism, there would be good rains.[17]

Mercury (Buddha)

It's a tiny planet, just larger the size of earth's moon. Like that of the moon, it, due to its smaller size, has no atmosphere to hold down gases and water. This is why the cratered planets; days are scorching hot and nights are frigid like that of the moon. Though it has all characteristics (qualities) like that of the moon, astronomically, it differs from the moon. Moon dominates the water element, so it is responsible for creating the phenomenon of rain. The Mercury, whereas, rules the element of air. Due to this reason, the planet is also responsible for giving effect mainly to the meteorological phenomenon of rainlessness or drought except for a few cases of rain. For instance, the Mercury while passing through the zones of *Śrāvaṇa*, *Dhaniṣ ṭhā*, *Rohiṇi*, *Mṛgaśirā*, *Uttarāṣāḍhā* cuts anyone of them across will give effect to the meteorological phenomenon of anti-rain.[18] When the Mercury enters the zones of the constellations of *Ārdrā*, *Punarvasu*, *Tiṣya*, *Āṣleṣā* and *Maghā*, it will create the conditions of drought.[19] On the other hand, while passing through the zones of *Bharaṇī*, *Kṛttikā*, *Rohiṇi* or *Svāit*, it causes good rain.[20] In its *saṁkṣ ipta* and *miśra* courses, i.e. while passing through *Puṣya*, *Punarvasu*, two *Phalgunis*, *Mṛgaśirā*, *Ārdrā*, *Maghā*, *Āśleṣā*, etc., it causes effects of mixed nature, i.e. sometimes rain and at other times drought.[21]

Jupiter (*Bṛhaspati*)

This is the largest planet in the Solar system with the turbulent cloud cover marked with an enormous eddy in them. Astronomically, it dominates fire and water elements. On account of its nature of always being covered with turbulent clouds and its domination of fire and water elements, it gives effect to the numerous towering clouds yielding sufficient rain whenever it comes closer to the earth.[22]

Its movement through various constellations creates various types of meteorological phenomena on the earth, on account of collusion of its basic character with those of various asterisms coming into its contact. For instance, if the Jupiter is eclipsed while moving to the zone of the 1st quarter of *Dhaniṣṭhā*, i.e. the tenth house and rises after eclipsing in the month of *Māgha* (which is represented by fire element), it will instead give rise to drought, the outbreak of fire's and storms.[23]

Should the Jupiter rise after the eclipse in the zone of *Mṛgaśirā* asterism (Ist two-quarters of which lie with the second house, i.e. the house of fire element and the last two quarters lie with the 3rd house, i.e. the house of air), it would likely to cause the rainlessness. This is because of the higher ratio of air and fire than water. Keeping in view the fact behind these statistics, the later meteorologists declared that should the Jupiter rise with *Mṛgaśirā* asterism, there would be drought.[24]

If the Jupiter rises in the zone of *Maghā* constellation (from the fifth house represented by fire element), it adds to a good rain. The same fact has been observed by Varāhamihira.[25]

Should it rise in the zone of *Phalgunis* (constituted of 5 portions of fire and three portions of air), its new constituent ratio of fire and air being 6:4, it will give effect to sporadic rain-fall.[26]

Its rise in the zone of *Citrā* constellation change its elementary composition as air 4, fire 9, water 9 and so it will be able to effect only scanty rain-fall. This is why, the later meteorologists maintained that the *Caitra* month have scanty rain-fall.[27]

With its rise in the zone of *Viśākhā*, the proportion of new composition works out as air 3, water 10, and fire 9 which is sufficient to bring fairly good rain for the growth of crops and so was also observed by ancient meteorologist Garga.

'In the month of *Vaiśākha*, rain, sufficient for the growth of crops, takes place.'[28]

In the zone of *Aṣāḍhā* constellation, the proportion of fire, earth and water being 14:3:9, it affects the phenomenon of rain at some places and rainlessness elsewhere.[29]

In the *Bhādrapadas* the revised statistics being air 3, water 13, and fire 9, it will give rise to rain sometimes and rainlessness at other times.[30]

In the *Aśvinī* constellation, the elementary proportion of fire and water being 13:9, it causes constant rain-fall because its power is boosted with the addition of four portions of fire.[31]

Jupiter's movement to the north of the constellation in the zone of which it rises was also observed by the ancient meteorologists as effecting the phenomenon of rain on the earth.[32]

Venus (*Śukra*)

Venus is earth's twin so far as its size and mass are concerned. It is also earth's next-door neighbour like that of Mars. It is seeringly hot and perpetually veiled behind the reflective sulphuric acid clouds. It has a carbon-dioxide environment over flat, rocky plains. The signs of volcanic activity have also been detected by the modern probe. Astronomically it is said that it rules the 2nd and 7th houses i.e. it dominates the element of water. Mythologically, it is known as the *Purohita* (precursor) of *asuras*, i.e. clouds. perhaps, this is why, during the governance of Venus, the year witnesses heavy rain poured out of the mountain like clouds. The ancient meteorologists also noted the same phenomenon.

Its various movements give effect to various types of weather conditions on the earth.

It has often been observed that its movement through anti-rain constellations brings about rain on the earth and through rain-producing constellations prevents rain on the earth.[34]

For instance, if the Venus passes through *Kṛttikā,*[35] *Ārdrā,*[36] *Puṣya,*[37] *Maghā,*[38] *Pūrvaphalgunī,*[39] *Uttaraphalagunī,*[40] *Swāti,*[41] *Viśākhā*[42] and

Purvabhādrapada,[43] it will cause heavy rain. On the contrary, its movement through *Anurādhā, Jyeṣṭhā, Mūla* and *Hasta* causes drought on the earth.[44]

Similarly, if the Venus is housed in the eastern horizon during the rainy season (monsoons) there will be scanty rain-fall. On the contrary, its presence in the western horizon causes copious rain.

Venus' rising or setting in different lunar days or *tithis* gives effect to different types of weather conditions on the earth. For instance, if it rises after an eclipse or gets eclipsed on the 8th, 14th or 15th (i.e. new moon day) or the dark fortnight, there will be heavy rain storms causing floods.[45]

If it is eclipsed or rises after the eclipse in the month of *Kārttika,* there will be no rain at all for the 90 days.[46]

Should the Venus appear or disappear at the time when the sun or the moon are completing their circuit, the rains would be effected.[47]

Saturn (Śanaiścara)

Saturn is the second-largest planet in our solar system. Its closeness to earth gives effect to the rainlessness.[48]

When it moves through the 'Front gates', *viz.* the seven asterisms commencing with *Kṛttikā* and becomes retrograde, there will be a dreadful and long-drawn famine and drought.[49]

Thus, we have studied above the weather-control on the earth by the individual planets passing through different constellations housed around the earth. Now we shall see the combined effect of the two or more planets moving through a particular house or houses.

The combined effect of two or more planets

If the combination of four or five planets takes place in the house of *Śvāti, Bharaṇi* and *Kṛttikā* constellations (astronomically known as *Nāgavīthi*), there will be heavy rain.[50]

If this combination occurs in the constellations of *Pūrvāṣāḍhā* and *Uttarāṣāḍhā* (popularly known as *Amalavīthi*), there will be no rain at all.[51]

When the sun, Mars and Saturn (known in astrology as malefic or *Pāpagrahas*) are in conjunction with *Kṛttikā* and *Rohiṇi* stars, there will be danger from fire and winds.[52]

Similarly, if this combination is in conjunction with the constellation of *Purvāṣāḍhā* and *Uttarāṣāḍhā*, there will be a threat of famine.[5]

If the Jupiter and Venus are situated in the Western and Eastern horizons respectively or vice versa and are housed in the 7th zodiac sign from each other i.e. 180^0 apart, the earth will not experience rain-fall.[54]

If at the time of dawn, Venus and Jupiter sojourn the eastern and western horizons respectively or vice-versa opposing each other, there will be rainlessness.[55]

Should the Jupiter, Mercury, Mars and Saturn precede Venus in their transit, there would originate hurricanes dashing trees, etc. to the ground thunder-bolt smashing the tops of mountains and not even a single drop of water would be there in the name of rain.[56]

When Mars should precede Venus in its transit, there would be complete rainlessness and lightning, fire and dust storm would be active.[57]

If the Jupiter takes precedence over Venus, hailstones will precipitate instead of rain.[58] But the precedence of Mercury over Venus gives effect to rain.[59]

Comets

In addition to the planets, comets, which originated at the time of origin of the planets and satellites, too play a vital role in affecting the atmospheric conditions on our planet. According to the investigations carried out by the ancient sages, there are 22 comets originated from the earth. They are round in shape like a mirror devoid of crest possessed of rays and are luminous like water or oil. They are housed in the north-east horizon. Their appearance portends famine.[60] According to another investigation, there are three comets originated from the moon, which is as white as the moon herself. They tenant the northern horizon. Their visibility brings about rain-fall.[61] Some other comets were also discovered that affected the atmosphere of the earth considerably. For instance, there is a big comet named *Vasātketu* which has elongated glossy body stretched out northward. It sojourns in the western horizon. Its rise brings about rain.[62]

There is another comet named *Asthiketu*, endowed with all the characteristics of *Vasātketu*, but is rough instead of glossy. Its appearance portends famine.[63]

There is yet another comet christened as *Jalaketu*, tenanting the western horizon and possessed of a crest and protruding towards west. Its rise gives effect to good rain.[64] One more comet called *Bhavaketu*, which is glossy, tiny and appears only for one night on the eastern horizon. It is endowed with a crest similar to that of a lion's tail curling to the right. It amounts to good rain for as many months as many *muhurtas* it remains visible.[65] The comet named *Kapālaketu* which is housed in the east and makes its appearance only on the *Amāvāśyā* / new moon day. Its rays and crest are of smoky colour. It traverses half of the horizon. It causes famine and drought.[66] The comet known as *Raudra* will also cause famine and drought-like that of *Kapālaketu*, if it rises in the east near *Purvāṣāḍhā* and *Uttarāṣāḍhā*. It has a tridental crest. Its flame is grey, rough and red. It traverses a third of the firmament.[67] The comet called *Calaketu* tenants in the western direction. It has a crest with its tip protruding south-ward. Its length increases with its advancement to the north. It makes its retreat after touching the seven sages, i.e. the Great Bear, the Pole star and the asterism *Abhijit*. It sets in the south traversing half of the firmament. Its appearance portends famine.[68]

There is a comet named *Śvetaketu*, whose crest is inclined southward and which sojourns in the east at midnight. There is another comet named *Ka* which has a yoke like a form and is housed in the west. Both can be seen at the same time for a period of 7 days. They are glossy and bring about rain leading to prosperity.[67]

The comet called *Kumuda* has a white lustre and a crest inclined towards east. It appears for a single night in the west. It bestows upon the countries it is sighted a unique prosperity caused by rain for a period of 10 years to come.[70]

The comet named *Maṇiketu* is tiny comet housed in the west and is visible only once for a period of three hours. Its crest is straight and white like a streak of milk ejected from the breast. As soon as it appears, it causes rain.[71]

Notes and References

1. *ākarṣaṇena rajasā vartamānaḥ.* *VS.* 33.43.
2. *tīkṣaṇaṁ tapatyaditijaḥ śiśire'pi kāla.* *BS.* 19.2.
3. *natyambudājalamuco'calasannikāśāḥ.* *BS.* 19.2.
4. See Origin of star and planets and Origin of Midsphere in author's book: *Origin of Universe.*
5. *vyāptaṁ nabhaḥ pracalitācala sannikāśaiḥ*
   vyālāñjanāliga valacchavibhiḥ payodaiḥ
   gāṁpūrayadbhirakhilāmamalābhiradbhi
   sutkaṇṭhitena guruṇādhvanitena cāśāḥ. *BS.* 19.4.
6. *yāte sthāṇudiśaṁ guṇāḥ subahavaḥ sasyārdhavṛṣṭyā dayaḥ.* *BS.* 24.33.
7. *spṛsannudagyātiyadā śaśāṅka stadā suvṛṣṭirbahulopasargā asaṅspṛśan yogamudak sametaḥ karoti vṛṣṭiṁ vipulāṁ śivam.* *BS.* 24.29.
8. *prāvṛṣi śitakaro bhṛguputrātsaptamarāśigatatah śubhadṛṣṭaḥ sūryesutannavapañcamago vā saptamagaśca jalāgamanāya.*

   BS. 28.19.

9. *śastra kṣudbhayakṛdyamena śaśijenā vṛṣṭidurbhikṣakṛt kṣemārogya subhikṣavināśī sītaṁśu śikhinā yadi bhinnaḥ.* *BS.* 4.27.

10. *abhyunnatā viyati saṁhatmūrtayo'pi muñcanti na kvacidapaḥ pracuraṁ payadāḥ.* *BS.*19.9.

11. *dakṣiṇato rohiṇyāścaram mahijo'rgha vṛṣṭi nigrahakṛt.* *BS.* 6.10.

12. *prājāpatye śravaṇe mūle triṣu cottareṣu śākre ca vicaran ghamnivahānāmupaghāta karaḥ kṣamātanayaḥ.* *BS.* 6.11.

13. *dvādaśa daśamaikādaśanakṣādvakrite kuje'śrumukham dūṣayati rasānudaye karoti rogānavṛṣṭiṁ ca.* *BS.* 6.2.

14. *asimuśalaṁ saptadaśādaṣṭādaśato'pi vā tadanuvakre dasyugaṇebhyaḥ pīḍāṁ karotyavṛṣṭiṁ saśastrabhayam.* *BS.* 65. *Parāśra* also records similar effects

saptadaśe'ṣṭādaśe vā dasyugaṇaiḥ prajānāmupadrava mvṛṣṭiṁ śastrabhayaṁ ca. (Bhaṭṭotpala on *BS,* 6.5).

15. *madhye na yadi maghānāṁ gatāgataṁ lohitaḥ karoti pāṇḍyo nṛpo vinaśyati śastrodyogādbhayamavṛṣṭiḥ.* *BS.*6.8.

16. *rudhirānanamiti vaktraṁ pañcadaśāt ṣoḍaśācca vinivṛtte tatkālaṁ mukharogaṁ sabhayaṁ ca subhikṣamāvahati.*

17. *vyālaṁ tryodaśarkṣāccaturdaśādvā vipaccyate' stamaye daṅṣṭri vyālamṛgebhyaḥ karoti pīḍāṁ subhikṣaṁ ca.* *BS.* 6.3.

18. *vicaran śravaṇa dhaniṣṭhāprajāpatyenduvaiśvadevāni mṛdnan himakara tanayaḥ krotyavṛṣṭiṁ sarogabhayām.* *BS.* 7.2.

19. *raudrādīni yadā pañca nakṣatrāṇindunandanaḥ bhinatti śastra durbhikṣa-vyādhibhiḥ piḍyate jagat*
Kaśyapa quoted by Bhaṭṭotpala on *BS.* 7.3.
Cf. also
raudrādīni maghāntāni upāśrite candraje prajāpiḍa śastranipata kṣudbhayarogānā vṛṣṭisantāpaiḥ. *BS.* 7.3.

20. *prākṛta gatāmārogyavṛṣṭi sasyapravṛddhayaḥ kṣemam* *BS.* 7.14.

21. *saṁkṣipta miśrayormiśram.* *BS.* 7.14.

22. *vividhairviyadunnataiḥ payodairvṛtanurvīṁ payasābhitarpa yadabhiḥ.*
Surarājguroḥśubhevarṣe bahusasyākṣitiruttamaraddhiryukto. *BS.* 19.15.

23. *ādyaṁ dhaniṣṭhāśamabhiprapanno māghe yadā yātyudayaṁ surejyaḥ.*
ṣaṣṭyābdapūravaḥ prabhavaḥ sanāmnā prapadyate bhūtahitastadābdaḥ
kvacittvvṛṣṭiḥpavanāgnikopaḥsantitayaḥśleṣmakṛtāścarogāḥ. *BS.* 8.27.28.

24. *saumye'bde'nā vṛṣṭiḥ.* *BS.* 8.4.

25. *pitṛpiyāparivṛddhirmāghe hārdiśca sarvebhūtānām ārogya vṛṣṭidhānyārdhasampado mitralābhaśca.* *BS.* 8.6.

26. *phalgunavarṣe vindyāt kvacitkvacitkṣemavṛṣṭi sasyānī.BS.* 8.7.

27. *caitre mandā vṛṣṭiḥ.* *BS.* 8.8.

28. *vaisākhe tu sasyajanmā vṛṣṭayaḥ sambhavanti hi.* *BS.* 8.9.

29. *kvacidavṛṣṭirnyatra.* *BS.* 8.11.

30. *bhādrapade-kvacitsubhikṣaṁ kvacicca bhayam.* *BS.* 8.13.

31. *āśvayuje'bde' jasraṁ patati jalam.* *BS.* 8.14.
See also Vṛddha Garga quoted by Bhaṭṭotpala on *BS.* 8.14
paryātasasyānnajalakṣemśāśvayujaḥśivaḥ

32. *udgāroary subhikṣakṣemakaro vākpatiścaaran bhānām.* *BS.* 8.15.

33. *śālīkṣumatyapi dharā dharaṇī dharābha dhāsādharojjhitapayaḥ paripūrṇavaprā.* *BS.* 19.16.

34. Cf. Parāśara on BS. 9.36 quoted by Bhaṭṭotpala
avārśake bhe vicaran yadi varṣati bhārgavaḥ
vārśakarkṣagato vyaktaṁ sodaśārcina varṣati.

35. *bhindangato'nalarkṣam kūlātikrāntva vārivā hābhiḥ.* *BS.* 4.24.

36. *ārdrāgatastu salilanikarakaraḥ.* *BS. 4.26.*

37. *puṣye puṣṭā vṛṣṭiḥ.* BS. 9.27.

38. *bhindan maghāṁmahāmātrā doṣakṛd bhūrivṛṣṭiḥ.* BS. 9.28.

39. *bhāgye ambunivaha mokṣāya.* BS. 9.29.

40. *āryamṇe saliladāyī.* BS. 9.29.

41. *citrāsthe śobhanā vṛṣṭiḥ.* BS. 9.30.

42. *svātau prabhūta vṛṣṭiḥ aindrāgn'pi suvṛṣṭiḥ.* BS. 19.31.

43. *karoti cosmin sitaḥ salilam.* BS. 19.34.

44. *triṣvapi caiteṣvanāvṛṣṭiḥ.* BS. 19.32.

haste piḍā jalasya ca nirodhaḥ. BS. 19.30.

45. *caturdaśīṁ pañcadaśīṁ tathāṣṭamīṁ*
tamisrapakṣasya tithim bhṛgoh sutaḥ
yadā varjed darśanamastameti vā
tadā mahī vārimayīva lakṣyati. BS. 9.36.
Cf. also Kāśyapa quoted by Bhaṭṭotpala on BS. 9.36.
kṛṣnapakṣe himāvasyā caturdaśyaṣṭamīṣu ca
udayaṁ bhārgavaḥ kuryāttadā vṛṣti pramuñcati.

46. *kārttike tu yadā māsi kurute'stamayodayau*
tadāhnāṁ navatiṁ pūrṇām devo bhuvi na varṣati
Paraśara on *BS.* 9.36.

47. *astodaye tu śukrasya yadi candradivākarau*
āvṛttimārgaṁ kurvāte tadā varṣati bhārgavaḥ.
Parāśara on *BS.* 9.36.

48. *sasyāni mandamabhi varṣati vṛtra śatrur*
varṣe divākarasutasya sadāpravṛtte. BS. 19.21.

49. *prāgdvāreṣu caran raviputro nakṣtreṣu karoti ca vakram*
durbhikṣaṁ kurutemahadugraṁ mitrāṇāṁ ca virodhamavṛṣṭim.
BS. 47.13.

50. *atijala mokṣaṁ kuryunāgākhyāyāṁ ca sarve tu.*
Samāsa saṁhitā quoted by Bhaṭṭotapala on *BS.* 20.9.

51. *durbhikṣa roga taskara śastrā vṛṣṭi kṣudhaṁ grahāḥ kuryuḥ*

Samāsa saṁhitā quoted by Bhaṭṭotapala on *BS.* 20.9.

52. *dehe krūra nipiḍite'gnyanilajaṁ* (*bhayam*) *BS.* 8.19.
See also Kāśyapa quoted by Bhaṭṭotpala on *BS.* 8.19.
kruragrahāhatedehedurbhikṣānalamārutāḥ.

53. *nābhyāṁ bhayaṁ kṣutkṛtam.* *BS.* 8.19.
See also Kāśyapa quoted by Bhaṭṭotpala on *BS.* 8.19.
kṣutbhayaṁ tu bhavennābhyām.

54. *gururbhṛguścāparapūrvakāṭhayoḥ*
parasparaṁ saptamarāśigau yadā
tadāprajārugbhayśokapiḍitā
navāripaśyantipurandarojjhitam. *BS.* 9.37

55. *pratyuṣe prāksthitaḥ śukraḥ pṛṣṭhataśca bṛhaspatiḥ yadā'nyo'nyaṁ nirīkṣete........avṛṣṭiśca tadā bhavet.*
yadā tu pṛṣṭhataḥ śukraḥ puracaśca bṛhaspatiḥ
yadā valokapetāṁ tau tāvadeva phalaṁ bhavet
Bhadrabāhu quoted by Bhaṭṭotpala on *BS.* 9.37.

56. *yadā sthitā jīvabudhārasūryajāḥ*
sitasya sarve'grapathānuvarttinaḥ
........bhavanti vātāścasamucchritāntakāḥ. *BS.* 9.38.
nacālpamapyambu dadāti vāsavo
bhinatti vajreṇa śirāṁsi bhūbhṛtām. *BS. 9.39.*

57. *nihanti śukraḥ kṣitije'grataḥ prajām*
hutāśaśastra kṣudvṛṣṭitaskaraiḥ
carācaraṁ vyaktamathottarapathaṁ
diśo'gnividyadrajasā ca piḍayet. *BS.* 9.41.

58. *bṛhaspatau hanti puraḥsthite sitaḥ*
site samastaṁ dvijagosurālayān
diśaṁ ca pūrvāṁ karkāsṛjo'mbudā
mlegadā bhūri bhavecca śārdam. *BS.* 9.42

59. *saumyo'stodayayoḥ purobhṛgusutasyāvasthitastoyakṛd.*
BS. 9.43.

60. *darpaṇavṛttākārā viśikhāḥ kiraṇanvitā dharātanayā*

kṣudbhayadā dvāviṁśatiraiśānyā mamabutailanibhāḥ.
BS. 11.13

Cf. also Garga quoted by Bhaṭṭotpala on *BS.* 11.13.
samastavṛttā viśikhā raśmibhiḥ parivāritāḥ
ambutaila pratikāśā dvāviṁsad bhūsutāḥ smṛtāh
aiśānyāṁ diśi dṛśyante durbhikṣabhayadāstute.

61. *śasikiraṇa rajatahimakumudakunda*
kusumopamāḥsutāḥ śaśinaḥ
uttarto dṛśyante tryaḥ subhikṣāvahāḥ śikhinaḥ. *BS.* 11.14.
Cf. also Garga quoted by Bhaṭṭotpala on *BS.* 11.14.
candraraśmisavarṇābhāhimakundendusaprabhāḥ
trayaste śaśinaḥ putrāḥ saumyāśāsthāḥ śubhāvahāḥ.

62. *udagāyato mahān snigdhamūrtiraparodayī vasāketuḥ*
sadya karoti marakaṁ subhikṣamopyuttamaṁ kurute.
BS. 11.29.

63. *tallakṣaṇa'sthi ketuḥ sa tu rukṣaḥ kṣudbhayāvahaḥ proktaḥ.*
BS. 11.30.

64. *Jalaketurapi ca paścātsnigdhaḥ śikharyāpareṇa cannatayā nava*
māsān sa subhikṣaṁ karoti śāntiṁ ca lokasya. *BS.* 11.46.

65. *bhavaketurekarātraṁ dṛśyaḥ prāk sūkṣmatārakaḥ snigdhaḥ*
harilāṅgulopamayā pradakṣiṇāvartayā śikhayā
yāvat muhūrtān darśanamāyāti nirdiśenmāsān
tāvadatulaṁ subhikṣaṁ rūkṣe prāṇāntikān rogā.
BS. 11.47,48.

66. *dṛśyo'māvasyāyāṁ kapālaketuḥ sadhūmaraśmiśikhaḥ*
prāṅnabhaso'rddha vicārī kṣunmarakā vṛṣṭirogakaraḥ.
BS. 11.31.

67. *prāgvaiśvānaramārgeśulāgraḥ syāvarūkṣatāmrārciḥ*
nabhasastribhā gagāmī raudra iti kapālatulyaphalaḥ.
BS. 11.32.

See also Parāśara quoted by Bhaṭṭotpala on *BS.* 11.32.
nabhastribhāgacārī sa śastrabhaya rogadurbhikṣa' nā
ṣṭimarakairyāvānmāsān dṛṣyatetāvadvarṣāṇitribhāgaśeṣāṇi
prajāṁ kṛtvārdhaṁ ca śāradadhānyamāḍhakamastaṁ vrajati.

Also Vṛddha Garga quoted by Bhaṭṭotpala on *BS.* 11.32.

tribhāgaṁ nabhaso gatvā tato gacchatyadarśanam
yāvato divasāṁstiṣṭhettāvadvarṣāṇi tad bhayam
śastrāgnibhayarogaiścadurbhikṣamaraṇairhatāḥ.

68. *aparasyāṁ calaketuḥ śikhayāyāmyāgrayaṅgulocchritayā*
gacchedyathā yathodak'tathā tathā dairghyamāyāti
saptamunīna saspṛśya dhruvamabhijitam eva ca prati nivṛttaḥ.
BS. 11.33; 34.

nabhaso'rddhamātramitvā yāmyenāstaṁ smupayāti
anyāmapi ca sa deśān kvacit kvaciddhanti rogadurbhikṣaiḥ.
BS. 11.36.

Cf. also Paraśara quoted by Bhoṭṭotpala on *BS.* 11.33-36.

teṣvapi kvacit kvacicchastra durbhikṣavyādhimarakabhayaiḥ kliśnātityaṣṭādaśa māsāniti.

Also Garga on *BS.* 11.33-36.

kṣucchastra marakayādhi bhayaiḥ sampiḍyet prajāḥ
māsān daśa tathāṣṭau ca calaketuḥ rudāruṇaḥ.

69. *snigdhau subhikṣaśivadāvathādhikaṁ dṛśyate ka nāmāyaḥ.*
BS. 11.38.

Cf. Paraśara quoted by Bhaṭṭotpala on *BS.* 11.38.

kaḥ prajapatiputro yadyadhikaṁ dṛśyate tadā dāruṇataraṁ
prajānāṁ śastrakopaṁ kuryāttathaiva snehavarṇayuktau
kṣemārogyasubhikṣadau bhavataḥ.

70. *kumuda iti kumudakāntivārunyāṁ prākśikhoniśāmekām*
dṛṣṭaḥ subhikṣamatulaṁ daśa kila varṣāṇi sa karoti.BS. 11.43.

71. *sakṛdekayāma dṛśyaḥ susūkṣam tā ro'pareṇa maṇiketuḥ*
ṛjvī sikhāsya śuklā stanodgatā kṣīradhāreva
uḍayanneva subhikṣaṁ caturo māsān karotyasau sarddhān.
BS. 11.45.

11

Astronomical Impressions Left by the Resultant Meteorological Phenomena on the Earth

The meteorological phenomena occurring in the atmosphere of the earth either due to geocentric formative factors or the extra-planetary controlling factors leaves various astronomical impressions or signs in terms of shape, appearance, colour, image, halo and size of the various heavenly bodies as well as of rainbows, twilights and clouds. In this chapter, a humble attempt will be made to assess the astronomical impressions left by the various meteorological phenomena.

Appearance

If humidity reaches the saturation point, it will make the astronomical phenomena appear glossy, lustrous, brilliant, dense or compact. For instance, glossy features of both the twilights signal the immediate rainfall.[1] If the sun-rays appear to by glossy, there will be rain.[2] Lustrous and glossy sun in the sky portends sufficient rainfall.[3] During the rainy season, the sun having a dazzling shining will effect rain within a short span of time.[4] Also, the sun bearing a shining like *śirīsa-puṣpa,* i.e. yellowish blue will effect rain shortly.[5]

The dense or compact moon also points to rain-fall.[6] The

moon with glossy and whitish rays is an omen of rain-fall.[7]

If the Saturn, only when it looks glossy, takes its course through *Śrāvaṇa, Svāti, Hasta, Ārdrā, Bharaṇi* and *Pūrvaphālgunī,* the earth will be inundated with waters.[8]

If the mountains wear the look like the heaps of collyrium or the caves wrapt in vapour, the rain will take place immediately.[9]

Colour

It has also been seen that meteorological phenomena also leave its astronomical impression in the form of colour. For instance, a saturated air of the atmosphere gives a blue colour to the star Canopus (*Agastya*).[10] It marks the white colour in the black *Rāhu.*[11] Sometimes humidity also provides the grey colour or yellowish shining to the *Rāhu.*[12]

The phenomenon of rain is also indicated by the yellow colour of lightning.[13] It makes the sun shine white, yellow or in a golden shade.[14] It attributes the sky with the white and blue colour or with the colour of cow's eyes or crow's eggs.[15] It attributes the moon with a red hue like the eyes of a parrot or a pigeon or of honey.[16]

On the contrary, the phenomena of anti-rain are marked by the multi-colours or the smoky colour of the sun-rays, or a black coloured sun in the rainy season.[17] Similarly, the white hue of lightning indicates famine conditions.[18] Red colour in *Rāhu* is also portentous of famine.[19] The meteorological phenomenon of anti-rain makes Venus appear grey like ashes.[20] It attributes the star Canopus with red or russet colour.[21]

Shape

The changing meteorological phenomena also give rise to the various astronomical shapes in the sky. For instance, the phenomenon of rain is marked by the moon with the northern peak looking up.[22] It also attributes the moon with the boat like shape with its northern peak having attained more shining.[23] The phenomenon of rain-fall also provides the moon with a shape of a tambourine.[24]

On the other hand, the meteorological phenomenon of anti-rain is marked with the sign of the form of a *kīlaka* in the sun.[25] Its impact on the moon reflects in the *Vajra* moon, i.e. the moon with a slender waist.[27] It also makes the moon sometime appear with the southern peak looking up.[28] Sometimes, it attributes the moon with a shape called *āvarjita,* i.e. the moon with the upward front.[29]

Size

Sometimes, the meteorological phenomena also tell upon the size of the various heavenly bodies. For instance, the phenomenon of rain-fall is indicated by the big size of the moon. On the other hand, the extremely small size of moon foretells the possible occurrence of famine.

Image

The phenomenon of rain-fall causes the development of the image of the sun to the north of it and also of the moon reflected in the sky.[30]

The phenomenon of anti-rain is reflected by the image of the sun developed to the south of it. Also, the image to both sides, i.e. north and south foretells the water crisis.[31]

Halo

The meteorological phenomena also leave their impression in the form of halos formed around various heavenly bodies. For instance, the halos of sun and moon that resemble the peacock's neck in colour are impressed by the phenomenon of rain.[32] Similarly, a thick and glossy halo that possesses the single colour assigned for the season and is covered with little razor-like clouds, or a yellow halo accompanied by the fierce sun is also formed to signal rain-fall. Moreover, all the haloes around Saturn and Mercury are the herald of rain.

A deep red coloured halo-like cock's eyes around the moon is also impressed by the saturated air.[33]

Rainbow

Rainbow is also one of the forms of astronomical impressions left by meteorological phenomena of rain. Rainbows are formed of various sizes, shapes and colours according to the various rainy conditions. For instance, if a rainbow is unbroken, bright, glossy thick multi-coloured and touching the earth at both extremities, and if it appears double, it is a sign of good rain.[34]

A rainbow seen in the east when there is no rain will produce rain and vice-versa; one seen in the west always indicates rain.[35]

Notes and References

1. *sarvairetaiḥ snigdhaiḥ sadyo varṣam.* *BS.* 47.22.
2. *snigdhā dīdhitayaī........vṛṣṭi syād.* *BS.* 47.23.
3. *ayane suprabhaḥ snigdhaḥ sevate yadi bhāskaraḥ.* *BS.* 3.5.
4. *prāvṛṭa kāle sadyaḥ karoti vimaladyutir vṛṣṭim.* *BS.* 3.28.
5. *varṣākāle vṛṣṭiṁ karoti sadyaḥ śīrīṣapuṣpābhaḥ.* *BS.* 3.27.
6. *sthūlaḥ subhikṣakārī.* *BS.* 4.20.
7. *singdhaḥ prasanno raśmivānśvetaḥ kṣemasubhikṣavṛṣṭikaraḥ.* *BS.* 4.30.
8. *sravaṇānilahastārdvā bharaṇī bhāgyopamaḥ suto'rkasya pracurasalilopagūdāṁ karoti dhātriṁ yadi snigdhaḥ.* *BS.* 10.1.
9. *girayo'ñjana curṇasannibhā yadi vā vāṣpaniruddha kandarāḥ.......vṛṣṭidāḥ* *BS.* 28.6.
10. *nilo'tivarṣāya* See Paraśara quoted by Bhaṭṭotpala on*BS.*12.22.
11. *śvete kṣema subhikṣam* *BS.* 5.53.
12. *ādhūmre kṣema subhikṣamādiśenmanda vṛṣṭiṁ ca* *BS.* 5.55. *vimalakamaṇi pītābho vaiśyadhvaṁso bhavet subhikṣāya.* *BS.* 5.57.
13. *pitābhavati sasyāya* *PM.* 2.3.13.
14. *varṣāsu śuklaśca.* *BS.* 3.23.

15. *gonetrābhyāṁviyadvimalādiśo........*
kākāṇḍābhaṁ yadā ca bhavennabhaīḥ BS. 28.4.

16. *śukakapota vilocanasannibho madhunibhaśca yadā*
himdī dhitiḥ........patati vāri tadā na cireṇa. *BS.* 28.11.

17. *........salilaṁ nāśu pātayati.* *BS.* 3.22.
varṣāsvasitaḥ karotyanāvṛṣṭim. *BS.* 3.26.

18. *durbhikṣāya sitā bhavet.* *PM.* 2.3.13.

19. *aruṇakiraṇānurupe durbhikṣāvṛṣṭayo*
Cf. also *pāṁśuvilohitarūpaḥ-bhavativṛṣṭeśca* *BS.* 5.55.

20 *patati na salilaṁ khāt bhasmarukṣā sitābhe* *BS.* 9.44.

21. *kapilastva vṛṣṭim māñjiṣṭha rāgaśadṛśaḥ kṣudham* *BS.* 12.19.

22. *proktasthānābhāvādudagaccaḥkṣemavṛddhivṛṣṭikaraḥ.* *BS* 4.16.

23. *arddhonnate ca lāṅgulamiti.....subhikṣaṁ ca* *BS.* 4.9.
Cf. also Vṛddha Garga quoted by Bhaṭṭotpala on *BS.* 4.9.
yadā samaḥ pratipadi nausthāyī sampradṛśyate
kṣema subhikṣamārogyaṁ sarvabhuteṣu nirdiśet.

24. *candro mṛdaṇgarūpaḥ kṣemasubhikṣāvaho bhavati* *BS.* 4.19.

25. *durbhikṣaṁ kīlake'rkasthe.* *BS.* 3.17.

26. *eko durbhikṣakaro* *BS.* 3.14.

27. *madhya tanurvajrākhyaḥ kṣudbhayadaḥ* *BS.* 4.19.

28. *dakṣiṇatuṅgaścandro durbhikṣabhayāya nirdiṣṭaḥ* *BS.* 4.16.
See also
yagameva yāmyakoṭyām kiñcituṅgaṁ sa
pārśvaśāyīti.......vṛṣṭeśca vinigrahaṁ kuryāt *BS.* 4.13.

29. *āvarjitamitya subhikṣakāri tad yodhanasyāpi* *BS.* 4.12.

30. *divaskṛtaḥ pratisūryyo jalakṛdudag* *BS.* 3.37.
pratiśaśī ca yadā divi rājate patati vāri tadā na cireṇa
BS. 28.11.

31. *dakṣiṇe sthite'nilakṛt ubhayasthaḥ salilabhayam* *BS.* 3.37.

32. *śikhigalasame'tivarṣaṁ........* *BS.* 34.6.

33. *kṛkavākuvilocanopamāḥ pariveṣāḥ śaśinaś ca vṛṣṭidaḥ.*

BS. 28.6.

34. *acchinnamavanigādhaṁ dyutimat*
snigdhaṁ ghanaṁ vividhavarṇam.
Dviruditamanulomaṁ ca praśastamabhbhaḥ
prayacchati ca *BS.* 35.3.

35. *vṛṣṭiṁ karotyavṛṣṭyāma vṛṣṭiṁ vṛṣṭyāṁ nivārayatyaindryāṁ*
paścātsadaiva vṛṣṭiṁ kuliśabhṛtaścāpamācaṣṭe. *BS.* 35.6.

12

Weather Forecast

In the Vedic times, the Vedic seer had total control over the weather. He was able to modify weather to the extent of his needs and requirements. That is why, he dared to proclaim as, 'Let the cloud precipitate as and when we desire.[1]' In the earliest Vedic phase, the science of *Mantra* was all the more practised by the high-spirited Vedic Ṛṣis at the level of their *Asaṁprajñāta Samādhi.* At this stage, the *Yoga* was the main tool or instrument for modifying weather. He could make it to rain, or ward off the same instantly at his will (*Mantra*). Later on, owing to the decline in his spiritual powers, he could practise *yoga* only up to the level of *Saṁprajñāta Samādhi* and at this stage, his means of weather modification also changed. Instead of *Mantra,* he could induce rain with the help of *Tantra* or the divine power achieved for rainmaking through *Saṁprajñāta yoga.* At the same time, keeping in view the decline of the tradition of *Brahmarṣis* (the seers practising *Asaṁprajñāta yoga* or *Brahma yajña*) and Ṛṣis (the seers practising *Asaṁprajñata yoga* or *Ātmayajña*), a more gross form of the means of the weather modification was also developed by the Vedic Ṛṣis which was christened as *Devayajña* or *Yajña.* The *Devayajña* form of *Yajña* was applied by the lay Ṛṣis in the Vedic times. Thus all the three forms of *Yajña* were the main tools or instruments of the Vedic *Ṛṣis* for weather modification. This fact was alluded to by the Vedic seer as:

'Make it to rain for me with the help of *Yajña*'. [2]

Thus weather was totally under the control of Vedic

seers. They could modify it to such an extent as it could suit their requirements and needs. As such, there was no need to think about the weather forecasting or weather predictions and so the Vedic seers didn't develop any means of weather forecasting. But by and by the people forgot all the three means of weather modification and thus they had to remain dependent upon the day to day weather for their needs and requirements. Instead of reviving the old Vedic techniques of weather modification, they carried out the in-depth studies to understand the behaviour of the atmosphere and change in the weather conditions, so that a reliable and efficient technique of weather forecast may be developed.

During the course, the change in weather conditions and the behaviour of the atmosphere was studied (1) in the context of the movements of various stars and planets; (2) in the context of physical process involved in the materialization of meteorological phenomena on the earth; (3) the astronomical impressions left by the resultant meteorological phenomena on the earth; (4) in the context of premonitions had by the various sensitive creatures on the planet; (5) and in the context of the representation of the entire season by one single day.

Based on the results of these studies five means/methods of weather forecast were evolved in the Vedic age which may be summarized as under:

Forecast on the basis of physical process

The Vedic Ṛṣis had already given a detailed break-up of the physical process involved in the materialization of various meteorological phenomena. Later on, when the science of artificial weather modification was forgotten, the knowledge of the same process was applied for making weather predictions.

Chapters four to seven discuss in detail the physical process involved in the formation of clouds, electrical charge, thunderstorms and precipitation.

Forecast on the basis of extra-planetary factors

The investigation of ancient meteorologists regarding the weather control on the earth led them to conclude that the movements of the heavenly bodies largely account for the creation of the various meteorological phenomena on the earth. The chapter eleven given above has already detailed the meteorological effects of the movement of heavenly bodies on the earth as studied by our ancient sages. Hence keeping in view the same fact, a new technique of weather forecast was developed on the basis of the extra-planetary factors.

Forecast on the basis of the Astronomical Impressions

The ancient meteorologists found that the various meteorological phenomena occurring in the atmosphere of the earth either due to the geocentric formative factors or the extra-planetary controlling factors leave various astronomical impression in terms of giving various shapes, sizes, appearances, colours, images, halos, etc. to the various heavenly bodies, rainbows, twilights and clouds (A detailed study of these astronomical impressions has already been done in chapter 12 above). As such, the ancient meteorologists made all these resultant astronomical impressions as their tools to forecast/predict weather in the days to come.

Forecast on the basis of Prognostics

During the course of studying the effect of meteorological phenomena in the context of astronomical as well as the human environment, it was found that the living creatures were also very sensitive and alive to the natural happenings or climatic conditions. They had the power to premonition the future weather conditions. Hence the behaviour of atmosphere/weather was also studied in the context of the special behaviour observed by the living creatures. A few investigations made in this direction by the ancient meteorologists may be cited as under.

For instance, much tumbling of fishes ashore and repeated croaking of frogs are prognosticative of rain-fall.[3] Immediate rain is also heralded by cats scratching the ground vehemently with their nails, and the construction of bridges on roads by children.[4] If the dogs weep standing on the roofs of houses or looking up towards the sky, the earth will be enlevelled with the water.[5] If the birds bath in water or sand and reptiles are seated on the tips of grass, there will fall rain immediately.[6] Should the ants shift their eggs without any trouble, snakes mat or climb down the trees, cows leap, the chameleons perch on the tips of trees and fix their gaze on firmament, and cows or (bulls) look up towards the sun, there would be rain ere long.[7] If the domestic animals like cows are reluctant to go out of the house and shake their ears and hoofs, or if the dogs behave in the same manner, it should be known that there will be rain soon.[8]

Forecast on the basis of One Single Day

Apart from the aforementioned inductive methods of the forecast, ancient sages also discovered a deductive method of the weather forecast. In course of investigating the various laws of nature, they found that the whole rainy season beginning from *Śrāvaṇa* to *Āśvin* is represented by the One Single Day, i.e. the day of moon's conjunction with the star *Rohiṇī.* The conjunction of the moon with the *Rohiṇī* takes place in the dark fortnight of the lunar month of *Āṣāḍha.* The seers like Garga, Parāśara, Kāśyapa and Maya investigated that the weather conditions to prevail in the entire rainy season may be predicted on the basis of weather conditions prevailed in one single day. Now the question arises as to how to get the prospects of weather of the entire season consisting of four months on the basis of one single day? The answer is very simple. The rainy season generally consists of four months commencing with the lunar month of *Śrāvaṇa.* These four months have eight fortnights and, on the other hand, a day consists of eight watches of three hours each. Hence, the eight watches of a day can be equated with the corresponding eight fortnights of a season, and its sub-divisions as days, i.e. twelve minutes of a day can be taken for one day of the rainy season and one hour for four days.

One should observe the behaviour of the wind throughout the day when the moon is tenanting the star *Rohiṇī* and predict the weather for the rest of the season accordingly.[9] Varāhamihira gives here some guidelines as to how to predict weather on the basis of the wind's behaviour. According to him, if the wind blows round from east to the south and so on, i.e. in a clock- wise manner, it is always indicative of good rain. When two contrary winds blow, the prediction should be made from the point of that wind which one is stronger.[10] So, it can be maintained at the end that the position and behaviour of the wind in each hour or each *prahara* (watch) of the day foretell the prospects of weather to prevail in each corresponding group of days or fortnights of the rainy season.

Thus from the foregoing discussion, it can be inferred that the ancient Indians, who were not able to modify the weather as per their needs and requirements, developed the techniques (mentioned above) of exact forecasting of short, medium and long ranges. They were able to predict the weather for more than six months through their efficient means of effective forecasting. They didn't develop any satellite system to get the synoptic view and computer facility to compute the statistical and numerical data taken for the purpose of weather forecasting. In other words, they were not merely puppets of these instruments, rather they were seers-the masters of vision. They observed the whole natural phenomena through *Yoga* and framed such natural laws as were capable of defining the various secrets of nature beyond time and space.

On the other hand, modern meteorologists have evolved three methods to understand and forecast the behaviour of the atmosphere. The important one being the synoptic technique wherein dominant weather producing systems are traced and evolution and movements are monitored. The advent of satellites has added new dimensions to this old method. The second and third method being statistical and numerical respectively. The invention of super-computers has also revolutionized these methods. With all these sophisticated instruments though the old techniques have been revolutionized to the extent of analyzing data hastily and accurately and getting the synoptic view without any flaw of subjective prejudice, but the drawbacks that lie in the very nature of these techniques cannot be removed. This is why, the modern meteorologists, though with the help of all the sophisticated devices are successful in short range forecasting, i.e. weather forecast up to 48 hours in advance. The medium (3 to 5 days) to long range (i.e. a month or more than a month) predictions are still largely a gamble.

Notes and References

1. *nikāme nikāme naḥ parjanyo varṣatu*
2. *vṛṣṭiśca me yajñena kalpatām.* *VS.* 18.9.
3. *jhaṣāḥ sthalagāmino rasanamaskṛṁmaṇḍukānāṁ jalāgama hetavaḥ.* *BS.* 28.4.
4. *marjārā bhṛśam avaniṁ nakhair likhanto rathyāyāṁ śisuracitāśca setubandhāḥ samprāptaṁ jalamacirannirvedayanti.* *BS.* 28.5.
5. *yadā sthitā gṛhapaṭaleṣu kukkurā rudanti vā yadi viyanmukhāḥ........kṣamā bhavati samaiva vāriṇā.* *BS.* 28.10.
6. *snāyante yadi jalapāṅsubhirvihaṁgāḥ sevante yadi ca sarīsṛpāstṛnāgrāṇya sanno bhavati tadā jalasya pātaḥ.* *BS.* 28.13.
7. *vinopaghātena pipīlikānāmaṇḍopasaṁkrānti rahivyavāyaḥ drumāvarohaśca bhujaṅgamānāṁ vṛṣṭernimittāni gavāṁ plutaṁ ca.* *BS.* 28.7.
   tarusikharopagatāḥ kṛkalāsā gaganatalasthitadṛṣṭinipātaḥ yadi ca gavāṁ ravivīkṣaṇamūrdhvaṁ nipatati vāri tadā na cireṇa. *BS.* 28.8.

8. *nechanti vinirgamaṁ gṛhāddhunvanti śravaṇān khurānapi*
paśavaḥ paśuvacca kukurā yadambhaḥ patatīti nirdiśet.
BS. 28.9.

9. *tatrārddhamāsāḥ praharair vikalpyā*
varṣānimittaṁ divasāstadaṁśaiḥ. *BS.* 24.10.

10. *savyena gacchañchubhadaḥ sadaiva*
yasminpratiṣṭhā balavān sa vāyuḥ. *BS.* 24.10.

PART -II

Experimental Meteorology

13

An Introduction to the Science of *Yajña*

Right from the beginning, *Yajña* was considered to be a strong and effective tool to get hold over the whole nature, internal as well as external. Various forms of this *Yajña,* such as *Ātmayajña* (self-accomplishment or *Yoga*), *Devayajña* (*Agnihotra* for augmenting the powers of elements of the physical world), *Pitṛyajña* (service to the elderly persons), *Atithiyajña* (service to the guests), *Balivaiśvadevayajña* (feeding other living creatures that surround us), *Brahmayajña* (doing the science of *mantras*) *Gṛhyayajña* (performance of various duties concerned with the compartmentalised life of an individual in the society), *Śrautayajña* (performance of symbolic rituals for the exposition of the process of creation enshrined in the Vedas), etc. were applied to attain accomplishment in various fields. *Ātmayajña* and *Devayajña* were the two ways or means of treating the various aspects of nature. *Ātmayajña* was an internal means applied particularly to treat the internal nature (*antaḥ prakṛti*). It was later known as *Yoga.* We find its full exposition in Patañjali's *Yoga-Darśana* or *Yoga*-Philosophy, according to which a *Yogī,* after attaining complete control over nature within becomes accomplished to have control over nature without i.e. external nature. Thus, in other words, it can safely be said that it was a spiritual way of treating nature. Actually, this spiritual/metaphysical aspect of Vedic science is more intriguing. This is what may be called an original contribution of the Vedic Ṛṣis. Modern science has so far not been able to realise the significance of this aspect. Since the spiritual method was very subtle one and so was within the reach of a few high-spirited Yogīs, its gross form was also developed simultaneously and named as *Devayajña* which was an external or material attempt to have control over natural powers. Contrary to *Yoga* which starts from within, it starts from without. Thus, the *Devayājī* is one who tries to control external nature first and accomplishes the self thereafter. The goal is the same, i.e. the realisation of the identity of the individual nature (self) with that of the universal one (God), but the ways of realisation are different.

One (*Yajña*) is the deductive method, i.e. it moves from universal nature to individual one. The other (*Yoga*) is the inductive one. It moves from individual nature to universal one. The actual attainment, however, is through the combination of both the *Yoga* and *Yajña.* In the modern times, tracking both the ways simultaneously Ram Narain Arya is going ahead with his experimentation on detecting the various inconsistencies and anomalies of the natural course like an able and dextrous physician does by feeling the nerve of patients and thereby treating the same with the help of *Yajña* in order to bring the nature down to suit the requirements of the living beings on the earth.

This is not going to be done for the first time in the history of mankind, but right from the Vedic period down, the *Yajña* was considered to be an effective tool to bring desired changes in the natural phenomena, e.g. to induce rain and to stop the same if the need is, to form clouds, to generate evaporation, to stimulate winds, etc. etc.

Thus the *Yajña* was the part and parcel of the Vedic life. In fact, it was the basis of human life. Everything of a Vedic Ṛṣi was abounding in *Yajña.* This may very well be understood through the perception of Vedic seer himself. He speaks thus:

yajñena yajñamayajanta devāstāni,

dharmāṇi prathamānyāsana

te ha nākaṁ mahimānaḥ sacanta

yatra purve sādhyāḥ santi devāḥ[1].

Another Ṛṣi proclaims:

yajñena yajñam ava yajanīyaḥ.[2]

Not only this, the *AV.* proposes to human beings to perform *Yajñas* season-wise daily, monthly and yearly, so that the seasonal deities may be stimulated to get their favour, e.g.

ṛtūn yajña ṛtupatīnārta vānuta hāyanān
samāḥ saṁvatsarān māsān bhūtasya pataye yaja.[3]

Keeping in view the importance of *Yajña,* authors of the *Brāhmaṇas* proclaimed it as the most noble act *yajño vai śreṣ ṭhatamaṁ karma.*[4]

Actually, the word *Yajña* was used in a very broad and wide sense of terms. It was derived from the root / *yaj* meaning *devapūjā, saṁgatikaraṇa* and *dāna. Devapūja* signifies the replenishment of all the natural forces existing in the midsphere and celestial sphere. The concept of *Devayajña* (*Agnihotra,* i.e. to augment the power of natural forces existing in the midsphere and celestial sphere) evolved out of the act of *Devapūjā* (replenishment work). *Dāna,* the offering made in kind to the living beings those are around us, gave rise to other forms of *Yajña.* For instance, the idea of *piṇḍadāna* (to offer provisions to the elderly persons of the society who are in a *vānaprastha* stage of life, to host a guest and to feed other creatures those live in our vicinity) helped evolve the concepts of *Pitṛyajña, Atithiyajña* and *Balivaiśvadevayajña* respectively. *Śaṅgatikaraṇa* is to create a new thing out of the combination or mixture of two or more things and hence the term *Yajña* also signified *Yāntrika* or technological development and advancement. Thus various forms of *Yajña* were evolved to attain accomplishment in various spheres of life. *Devayajña* or *Haviryajña* (which henceforth will be termed as *Yajña*) was the only means to do away with the physical world. As it has already been stated above that it (*Yajña*) was used as a strong and effective tool to bring the desired changes in the natural phenomena or say weather modification like the change of the airflow in terms of direction and speed, the formation of clouds, seeding supercooled clouds, rainmaking whenever and wherever so desired, preventing undesired rains, control of pollution and prevention of various diseases of men, animals and plants.

Theory and Principles of Yajña

Now the questions arise as to what are the basic principles underlying the science of *Yajña?* How does it matter in weather modification, pollution control and prevention of diseases? Before to get an answer to all these questions, it is necessary to understand the actual process of *Yajña.* Normally, a *Yajña* consists of the three elements: (i) *Agni* (ii) *āhuti-dravya,* i.e. an offering material (iii) and *devatā,* i.e. the deity to whom the material is offered. *Kātyāyana Śrautasūtra* speaks of *Yajña* as *dravyaṁ devatā tyāgaḥ.*[5]

> "I.e. the *Yajña* is offering the specific material to the specifically targeted deity by means of fire."

Thus all the above-mentioned elements, *viz.* fire, offering material and a deity are included in the aforesaid definition of *Yajña.* This is what is the actual meaning, purport and purpose of the *deva-yajña* form of the Vedic *Yajña.* All three elements play a vital and crucial role in the proper prosecution of *Yajña* mechanism.

Before we proceed further and take up all these elements one by one for consideration, it is necessary to be acquainted with the following Vedic principles in order to have a better understanding of the mechanism of *Yajña.*

Theory of Physical Embodiment of Universe

According to this theory, *yad aṇḍe tad brahmāṇḍe,* i.e. whatever is contained in a physical body is also embodied in the universe outside. In other words, there is a close similarity between an organic whole and the organisation of the universal whole. This similarity is also known as the similarity of *vyaṣṭi-śarīra* (individual's body) to the *samaṣṭi-śarīra* (universal body); so as the similarity of *puruṣa śarīra* to the *prakṛti śarīra*. Just as the universe is a composition of five gross elements, so is an organic body composed of these five elements.

These elements are supposed to exist in the organic body or universe in a balanced or organised manner or ratio. As long as this balance or organisation is maintained, the organic body survive and nature takes its normal course, but the imbalance or disorganisation of elements disturb the normal course of bodies and nature as well and ultimately leads to their destruction or end. Just as the air, fire and water are the dominant factors in the operation of the universe, so are the *vāta, pita* and *kapha* represented by the air, fire and water respectively in the organic body. Just as an able and dexterous physician feels the nerve of a patient to detect the weakness/illness caused by the imbalance of any of the three factors and treat him accordingly, similarly the inconsistencies in the matter of the natural course caused by the imbalance of the triad can be detected by a meteorologist and the same (imbalance) may be remedied after proper and suitable treatment of elements. Imbalance of *vāta, pita* and *kapha* in a physical body is treated by administering medicines to the patients, similarly, the remedy of the sick environment is also possible through medication is done with the help of *Yajña,* the method of which will be explained duly in the ensuing pages.

Theory of Quintuplication

According to the Vedic philosophy, *Īśvara* or God is the efficient cause of the creation and *Jīva,* the organic whole, and the *prakṛti* (matter) are the simple and material causes respectively. *Prakṛti* consists in the equilibrium of *sattva, rajas* and *tamas* qualities and the disequilibrium of these leads to the *vikṛti* or evolution which, among others, consists of five subtle elements or *sūkṣmabhūtas* known as *tanmātras* (net elements), *viz. ākāśa* (ether) *vāyu* (air), *agni* (fire), *āpaḥ* (waters) and *pṛthivī* (earth). At the initial stages, these elements being in their own atomic form are called subtle elements (*sūkṣmabhūtas* or *tanmātras*). All the five have their own distinct sensory properties, such as *ākāśa's* sensory property is sound, air's touch; fire's colour; water's taste and earth's smell.

These elements have a particular order or sequence of evolution. The element that comes first in order during the course of evolution causes the next one to follow. The preceding will be lighter and subtler than the following and just reverse of it will be the order and position during the course of dissolution. The process of evolution of five subtle elements into gross elements takes place owing to the fusion of the atoms of these elements which is known as the theory of quintuplication or *Pañcikaraṇa.*[6] Due to this process of fusion *ākāśa* (ether) was created first.

tasmāt vā etasmād ātmana ākāśa sambhūtaḥ,[7]

'That is from Brahman arose *ākāśa*'.

Ākāśa or ether was qualified by sound. Sound cannot exist without *ākāśa.* With the further process of ongoing fusion in the atoms, air came into being.

ākāśād vāyuḥ,[8]

'From *ākāśa* arose air'.

It had its own quality, the touch and also inherited the quality of ether, the sound. So, it had two qualities: sound as well as touch. Being the second one in order of evolution, it assumed a grosser form than the ether.

This is why it is present in the gaseous state and can only be known with the help of the skin sensation. Since air affected as caused by *ākāśa,* it cannot exist without *ākāśa.* Where there is *ākāśa* there is air and where there is air, there is touch as well as sound. The third one to come in order was fire.

vāyoragniḥ,[9]

'From air arose fire'.

Being the third one in the series, it assumed more grossness as compared even to air i.e. it got an ionized state or say a thick gas marked with a colour. Its own property is *rūpa,* due to which it becomes visible. In addition, sound and touch were inherited from the predecessors *ākāśa* and air. Since the fire owes its effect to air, it cannot exist without air. Where there is fire, there is air and where there is no air, there cannot be the fire.

The elements of air, *viz.* oxygen, which is helpful in burning, and hydrogen, which has a burning characteristic, are the main factors in the origin of the fire. The fourth one to come into effect is water.

agnerāpaḥ,[10]

'The fire gave effect to waters'.

It had its own property as taste and the sound, touch and colour were inherited from ether, air and fire respectively. It occupied a liquid state. It was caused by fire, hence it is composed of the fire agents, *viz.* oxygen and hydrogen. The last one to follow was the earth. It arose from waters-*adbhyaḥ pṛthivī.*[11]

It was still grosser than waters, hence got a solid form. Its own property is smell plus four of the preceding ones. It was an outcome of waters, hence still sustained by waters. Perhaps this is why the proportion of hydrosphere is much larger than lithosphere (i.e. 3:1).

In fact, all these five gross elements are the pointers to the five states of a matter: 1. ethereal state; 2. gaseous state; 3. ionized states; 4. liquid state; and 5. solid-state in descending order or in the order of their origin.

Territorial Division and Function of Deities

Devatā or deity is the name of a natural element or sub-element existing in various zones of the atmosphere. Actually, *Devatā* is the *deva* itself; *deva eva devatā* since like that of *deva* it also helps the existence and regulation of life on the earth by providing some necessity or other.

According to Yāska, an ancient Indian etymologist, *devo kasmāt? Devo dānātvā dyotanāt vā dyusthāno bhavatiti vā*[12] i.e. they are called deities, since (1) either they provide us with some necessities of life or help the existence of life cycle on the earth, e.g. the sun provides us with the light and sunshine, the *Mitra prāṇa* (oxygen), *Indra* lightning, *Soma* coldness and humidity in the atmosphere, *Marutas* air, *Varuṇa* hotness and dryness, *parjanya* clouds, *Abhra* super-cooled clouds, *Purūravā* thundering clouds, *Rudra* thunderstorms, *Aditi* vaporisation of terrestrial waters and other aqueous substances into the upper layers of the atmosphere, etc. (2) or since some of them are luminary bodies (3) or since some of them are celestial bodies or they have their abode in the *dyuloka* or firmament.

Actually, each natural phenomenon was considered to be the function of some unforeseen natural powers, and hence each one (natural phenomenon) was associated with some foster deity (*abhimāni devatā*). Before a natural phenomenon materializes, several other functions are required to work out. This diversity or variety of function in a natural phenomenon led to the diverse naming of the one and the same deity. Yāska also related this fact, as *tāsāṁ mahābhāgyādekai-kasya'pi bahūni nāmadheyāni bhavanti. api vā karma pṛthaktvāt.*[13] Śaunaka, the author of *Bṛhaddevatā*, bears out the same fact as:

etāsāmeva mahātmyān nāmānyatvaṁ vidhiyate
tattatshthāna vibhāgena tatra tatreha dṛśyate.[14]

See also

tāsāmiyaṁ vibhūtirhi nāmāni yadanekaśaḥ.
āhustāsāṁ tu mantreṣu kavayo' nyonyayonitām.[15]

These deities have been numbered sometimes 33, sometimes 3300, sometimes 33 lakh and sometimes 33 crores, according to their various functions and powers. The whole observable universe has also been divided into three zones, namely *bhūloka* (terrestrial sphere), *antarikṣaloka* (mid-sphere) and *dyuloka* (celestial sphere), each one is the abode of some gross element or the other and governed by a presiding deity and several others as subordinate deities. The terrestrial sphere is the main abode of solid and liquid matters such as earth and water and is governed by *Agni* as presiding deity and *Jātavedas, Vaiśvānara, Draviṇodā, Tanunapāt Nārāsaṁsa, Ilā, Vṛṣabha,* etc. etc. as subordinate deities.[16] The *antarikṣa loka* (mid-sphere) is the abode of gaseous elements, such as air and water vapours. It is governed by *Indra* and *Vāyu* as presiding deities and *Parjanya, Rudra, Bṛhaspati, Apāṁnapāt, Pururavā, Aditiḥ, Tvaṣṭā, Savitā, Vāta, Vācaspati, Soma, Marut, Aṅgirasa, Ribhu, Pitara,* etc. etc. as subordinate deities.[17] The *dyuloka* (celestial sphere) is the main abode of thermal elements and is dominated by *Sūrya* as a presiding deity and *Aśvinau, Vṛṣ ākapāyī, Sūryā, Uṣā, Pūṣā, Vṛṣākapi, Yama, Vaiśvānara, Viṣṇu, Varuṇa, Saptarṣi, Āditya, Savitā,* etc. as subordinate deities.[18]

Law of Specific Elementary Gravity / Similarity

As it has already been stated that all the things (living or non-living) in this world are composed of any one two or more of the five gross elements or say standard elements. This is why they are said to be the material cause (*upādāna kāraṇa*) of the whole physical creation. In other words, it can be said that all the things of this physical world are the *vikāra*, i.e. material forms or part of anyone two or more of the five standard elements. Thus these material forms/parts are the representatives of their respective standard elements. For instance, the terrestrial fire is the material form/part (*vikāra*) of the standard element, the celestial fire, and the latent heats such as *jaṭharāgni* (heat of stomach), *dāvānala* (fire caused by forest conflagration), *vaḍvānala* (submarine fire) and the acid fire are also the forms/part of the celestial fire. Similarly, ice and vapours are the forms/parts of the standard element, the water and the aqueous substances contained in solids and other liquid substances also represent the water. There always exists an elementary or material similarity between the standard elements and the forms/parts thereof. This elementary similarity gives an effect to the specific gravitational force between the standard element and the representative form thereof. Due to this specific gravitational force, a form/part always gravitates towards its respective standard element, no matter where it is abiding, whether in *bhūloka* (terrestrial sphere), mid-sphere or celestial sphere. Hence in the natural course of events, if there is no opposing factor, due to the law of specific elementary gravity, the form/part (*vikāra*) will move towards its respective standard element and will finally get mixed up with it and thus covers up its deficient/lost potential. For instance, when a human body dies, it is said, *pañcatvaṁ gataḥ*, i.e. the bodily substances composed variously of five standard elements mixed up with their respective elements. Patañjali, the author of the Mahābhāṣ ya, minutely observed this fact while laying down the principles of replacement of some sounds by the other on the basis of the similarity of their place of articulation in the context of *pāṇinian Sūtra: sthāne antartamaḥ*-1.1.50. He maintains thus:

> *tadyathā samājeṣu samāśeṣu samavāyesu cāsyatāmityukte. naiva kṛṣāḥ kṛśaiḥ sahāsate na pāṇḍavāḥ pāṇḍubhiḥ. yeṣāmeva kiñcidarthakṛtamāntaryaṁ taireva sahāsate. Tathā gāvo divasaṁ caritvatyo yo yasyāḥ prasavo bhavati tena saha śerate. yathā yathetānigoyuktakāni saṁghuṣṭ akāni bhavanti tānyanyonyamapaśyanti śabdam kurvanti. evaṁtāvaccetanāvatsu.*
> *aceṭaneṣvapi taḍyathā loṣṭaḥ kṣiptaḥ bahuvegaṁ gatvā naiva tiryag gacchati nordhavamārohati pṛthivīvikāraḥ pṛthivīmeva gacchatyān-taryataḥ. Tathā ca ya etā āntarikṣya sūkṣmā āpaḥ ākāśe nivāte naiva tiryaggacchati nārvāgāvarohati. ab vikaro'pa eva gacchatyāntaryataḥ. Tathā jyotiṣo vikāro'rcirākāśadeśe nivāte suprajvalitaṁ naiva tiryaggachhati nārvāgāvarohati. joytiṣo vikāro jyotireva gacchatyāntaryataḥ.*

'That is for example, at the time of congregations, assemblies or community dinners when the people are asked to take their seats, neither physically weak person will sit together with the physically weak nor will the patients of tuberculosis with the patients of tuberculosis, but those having similar status in terms of rank, property and profession will sit together. So far as animals are concerned, kines having grazed throughout the day will come to sleep along with their young ones. Similarly, the oxen, which were yoked together to plough the field, produces sound together at night even without seeing each other. These are the examples of union among the living beings on the basis of some or other type of similarity.

Likewise, the unification or for gathering takes place among material things on the basis of their elementary or material similarity. For instance, a lump of clay when thrown upward at a high velocity neither moves on curving path nor does experience an upward acceleration, but being the *vikāra* (form/part) of the earth moves downward to the earth. And the smoke emitted by a solid substance on being burnt is due to the aqueous element contained in the matter. Hence, it is the *vikāra* (form/part) of gaseous water abiding in the mid-sphere. That being so, instead of moving on the curving path or drifting downward, it experiences an upward movement to join the mid spatial water when the air remains suspended. So also does the terrestrial fire. It being the *vikāra* of celestial fire, neither takes a transverse path nor does move downward, but gravitated upward to its standard element (due to the specific gravitational force caused by the elementary similarity), when it is not opposed by the air acceleration.

Thus from the foregoing discussion, two things are clear.

The standard elements abiding in their particular spheres always maintain their link with their respective form/part residing in other spheres and gravitate them towards them.

The things composed of various standard elements will ultimately join their respective standard elements when they occupy the atomic state due to decay, destruction or by any other method. Thus they remedy the deficient/lost potential of that very standard element or deity.

Elements of Yajña

After giving a detailed break-up of the essential law factors that are fundamentals of the *Yajña*-mechanism, we shall discuss hereunder the three elements of *Yajña*, *viz.* *āhuti-dravya*, fire and *devatā* for the proper understanding of this phenomenon.

Āhuti-Dravya

Āhuti-dravya or an offering (oblation) material is a particular type of select material qualified to augment the material power or energy of the respective deity or *devatā* (being the form or part thereof) it is meant for. This particular type of offering material when released to the sacrificial fire is taken up by it in conjunction with the air in its gaseous form through its column established with the help of the gravitational force of celestial fire to the targeted deity in the upper layers of the atmosphere. There it mixes up with the targeted deity and thus remedies the deficient potential or say augments the power and property of the targeted deity (the natural element) by tens of thousands times in its spherical zone. The Veda also tells us that the *āhuti* of a particular substance is capable of augmenting the power and property of a particularly concerned deity and not of all the deities. For instance, the following statement of *Śatapatha Brāhmaṇa* may be noted:

> *taddhaike devebhyaḥ śundhadhvaṁ. devebhyaḥ śundhadhvamiti phalīkurvanti tadu tathā na kuryyādāaṣṭiṁ athaitadvaiśvadevaṁ karoti yadāha devebhyaḥ śundhadhvam iti vā etaddevatāyai havirbhavati.*

According to this statement of *Śatpatha Brāmaṇa*, 'Some experts pronounce *devebhyaḥ śundhadhavaṁ* while offering *āhuti*, but this should not be pronounced as the *āhuti* becomes *vaiśvadevī* (i.e. pertaining to all the deities) when he says '*devebhyaḥ śundhadhavam.*' Actually, a particular *āhuti* always pertains to a particular deity.

Devatā

Devatā or deity is the name of a standard element or sub-element abiding in the mid-sphere or celestial sphere to whom the *āhuti-dravya* is offered with the help of the *Yajñiya Agni* (sacrificial fire). The detailed description of the concept of the *devatā* has already been given above under the head: Terrestrial division and function of deities. Actually, natural elements or deities, such as *Soma, Varuṇa, Indra, Parjanya, Marutas,* etc. etc. lose their potential energy by and by in the natural course. Until and unless this deficiency is remedied, they won't be able to regulate the life cycle on their respective part. Under the circumstances, *Yajña*, the surface burner, is the only means with the help of which the deficient potential of these deities may be remedied by supplying necessary energy boosters through its mechanism. As it is said in the *Veda* itself.

Yajña devānāṁ pratyeti,[19] i.e. *Yajña* transmits provision to the deities. At another place the *Ṛgvedic* Ṛṣi wishes the *Yajamāna* that may this *Yajña* augment or boost the power/potential energy of the deity, *Indra.*

yajña hi te Indra vardhano bhūt.[20]

A yet another verse-*yajñena vardhat jātvedasam* 'may the *Yajña* augment the power of *Jātvedas*' also bears out the above-maintained fact.

Perhaps, owing to the same characteristic of *Yajña*, the Veda has loudly proclaimed to strengthen the essence or potential energy of each and every deity with the help of *Yajña* for the prosperity and well-being of the planet and planetary life. The verses read as follows:

devaṁ-devam avardhayat[21]
devaṁ-devaṁ yajñāmahe.[22]
devaṁ-devaṁ huvema vājasātaye.[23]

Fire

The terrestrial fire which is the representative form of the celestial fire has been attached great importance among the three elements of *Yajña.* As the *RV.* puts it: *agniṁ yajñeṣu pūrvyam.*[24] '*Agni* should be brought first in the *Yajñas.*' Yāska also by way of etymologizing expresses similar views regarding *Agni.* According to him, *agniḥ kasmāt agraṁ yajñeṣu nīyate*[25], i.e. *agni* is what is brought first in the *Yajñas.*

The *Ṛgveda* at the very outset calls *Agni* as the leader of the *Yajña* phenomenon, e.g.

agnimiḍe purohitam yajñasya devaṁ ṛtvijam hotāraṁ ratnadhātmam.[26]

This characteristic of *Agni* has been extolled at several other places. For example:

yajñasya ketuṁ prathamaṁ purohitamagniṁ[27]

In the Ṛgvedic verse 1.1.1. quoted above, it has been called *ṛtvija,* since the *Yajña* is performed with the help of agni *ṛtu*-wise (seasonally).

This characteristic of *Agni* has been extolled at several other places. For example:

yajñasya ketuṁ prathamaṁ purohitamagniṁ[27]

In the Ṛgvedic verse 1.1.1. quoted above, it has been called *ṛtvija,* since the *Yajña* is performed with the help of it *ṛtu*-wise (seasonally).

Characteristics of fire

Agni has three conspicuous characteristics:

Agni as an expanding agent

Generally, the solid or liquid substances are composed of the gross elements either of earth or water or of both. *Agni* effects or brings about the decomposition of things, solid or liquid, into parts of gross elements which they are made up of when treated with it. Similarly, an *āhuti-dravya* on being treated with fire is decomposed into its earthen parts as carbon and aqueous part as vapour, the combination of both is often called smoke.

The aqueous part of *āhuti,* i.e. the vapour being in the gaseous state gets defused with the air faster than the *āhuti* in a solid or liquid state and scatter in a wide range of area in the company of air thus expanding in terms of volume. On account of effecting this expansion of material offered to fire, *Agni* is extolled by the *Ṛgvedic* Ṛṣi as an expanding agent of *havis* or offering material thus, *agniṁ yajadhvaṁ haviṣā tanā girā,*[28] i.e. let the *yajña* be performed with the help of *agni* that expands the *havis.*

Agni as an oral agent of deities

Agni is often called as the oral agent of all the other deities abiding in the midsphere and celestial sphere of the universe. For instance:

agnirvaidevānāṁ mukham,[29]

i.e. *Agni* is the mouth of the deities.

mukhaṁ vā agnirdevatānām.[30]

Agnirmukhaṁprathamo devānām.[31]

'The terrestrial fire was the first to become the mouth of the deities.'

Agnir mukham.[32]

'*Agni* is the mouth.'

All these lines stand to signify that all the deities or natural elements receive their provisions through *Agni* as their mouth. Just as various limbs of the body receive their body supplies through the mouth, so do all the deities of midsphere and celestial sphere derives their provisions through their mouth, the *Agni.* Actually, this process is accomplished by way of *āhutis* released to the fire of *Yajña.* Perhaps this is why, the *āhutis* are called the soul of *Yajña, haviṁṣi vā ātmā yajñasya,*[33] and the *Yajña* are introduced as the provision of the deities, *yajña u devānām annam.*[34] In this context, *Maitrāyaṇi Saṁhitā* very aptly remarks that the deities or natural elements live on *Yajña-yajñamasya devā upajīvanti.* [35]

Agni as a transporting agent

As it has already been said, *Agni,* the terrestrial fire, is the form or part (*vikāra*) of the celestial fire. The Veda also points out to the same fact as:

divi mūrdhānaṁ dadhiṣe svarṣāṁ jihvāmāgne,[36]

'The *Agni* has the sun as its head in the celestial sphere.'

Hence, due to the law of specific elementary gravity, it always gravitates upward towards its standard element.

In this process, a fire column is established with the help of the gravitational force of its standard element. The air close to the fire column also gets heated up and hence accelerated upward. In fact, when the molecules/atoms of the smoke of the oblation (*āhuti dravya)* gets diffused with the molecules of this upward moving air, they reach in the upper layers of atmosphere and gravitate towards their respective deity or natural element. This phenomenon is very well observed by a Vedic seer. According to him,

svāhākṛte ūrdhavaṁ nabhasaṁ mārutaṁ gachhatam,[37]

I.e. on being sacrificed, an oblation (*āhuti)* becomes *māruti* (pertaining to the deity *marut*) and hence gets an upward thrust due to the upward rising air.

Due to the characteristics of transporting oblations (*āhutis)* to the upper layers of the atmosphere, *Agni* is evoked with such attributive epithets as *havyavāhanam,*[38] *kavyavāhana*[39] (transporter of *āhutis*) *pathikṛt* (path-maker of *āhutis*), *dūta*[40] (*āhuti* courier), etc. etc.

This process of transporting the oblation to the deities is accomplished in the company of air. This is why *Agni* is eulogized as

marutvānagnaye marutvate svāhā.[41]

In view of this characteristic of the fire element the Vedic Ṛṣi laid an emphasis on the growth or expansion of the fire of *Yajña* by way of various oblatory material (*āhuti-dravyas)* and *samidhās.* As for example,

agniṁ vo vardhantaṁ adhvarāṇām.[42]

agnim̐ haviṣā vardhantaḥ.[43]
agnir havirajuṣat.[44]
agnim̐ yaja.[45]
agnaye samidham āhārṣam.[46]
agnayesamidhyamānāyanubrūhi.[47]
agni ghṛtena vāvṛdhuḥ.[48]
ayam̐ yajño vardhatām̐ gobhiraśvaiḥ.[49]
ghṛtena tvam̐ jātavedaḥ tanuvo vardhayasva.[50]

This characteristic of *Agni* has been extolled at several other places.

To sum up, it can be said that *Agni,* the terrestrial fire, is the only agent which is capable of sending properly the particular oblations *(āhutis)* to their respective deities. This is why, Manu, the first lawgiver of humankind, says at one place:

agnāv prastāhutiḥ samyag ādityamupatiṣṭhati,[51]

I.e. the oblation *(āhuti)* released to the fire goes properly to the sun.

According to the authors of the *Brāhmaṇas, āhutis* are never offered without fire, (*na vā anagnāv āhutir hūyate*) because the *āhutis* offered without fire don't go to their respective deities.

na ha vā tā āhutayo devāngacchanti yā avaṣaṭkṛtā vā' svāhakṛtā bhavanti.[52]

Notes and References

1. *RV.* 1.164.50; 10.90.16; *AV.* 7.5.1.; *VS.* 31.16; *TS.* 3.5.11.5; *T.Ār.* 3.2.7.
2. *RV.* 3.32.12.
3. *AV.* 3.10.9.
4. *Ś. Br.* 1.7.1.5.; *Kāth.* 3.10.
5. 1.2
6. The process of *Pañcikaraṇa* proceeds as follows:

dvidhā vidhāya caikaikaṁ caturdhā prathamaṁ punaḥ
svasvetara dvitiyaṁśairyojanāt pañca pañca te

Vedāntasāra 28.

That is to say, each subtle element is broken into two halves. The second half is further broken into four quarters (i.e. one eight parts). Thus the first half of each gross element is constituted by half of its own atoms, the other second half of each gross element is constituted by the four quarters (1/8 parts) of an atom of each of the other four types of subtle elements. (e.g. gross *ākāśa* is one-half a subtle *ākāśa,* with the other half composed of 1/8 parts of air, fire, water and earth) and so on.

7. *Tait. Up. Brahmavalli, Ist Anuvāka.*

8. *Ibid.*

9. *Ibid.*

10. *Ibid.*

11. *Ibid.*

12. *Nirukta*

13. *Ibid.* 7.5.

14. *Bṛhaddevatā,* 1.70.

15. *Ibid.* 1.71.

16. See *agniḥ prithivī sthānaḥ-* (*Nir.* 7.5)
athaitānyagnibhaktīni. ayam lokaḥ prāthsavanam,
vasantaḥ, gāyatrī, trivṛtstoma, rathantaram sāma,
ye ca devagaṇāḥ samāmnātāḥ prathame sthāne. *Nir.*7.8

See also *Bṛhaddevatā* 1.73-*agni'smin.*

Also
yadyatra prithivīsthānaṁ pārthivaṁ cāgnimāṣritam
tadasarvamanu pūrvyeṇa kathyamānaṁ nibodhat.
(*Bṛd. 1.105.* Cf. also 1.106-1.120).

17. As to the principal deities, see *Nirutka* (7.5)

vāyuvendrovāntarik-ṣasthānaḥ.

And *Bṛhaddevatā* (7.73): *athendrastu madhyato vāyureva ca.*

As to co-deities see *Nir.* (7.10):

athetānīndra bhaktīni antarikṣalokaḥ, mādhyandinaṁ savanam, grīṣmaḥ, triṣṭup, pañcadaśastomaḥ, bṛhatsām, ye devagaṇāh samāmnātā madhyame sthāne yāśca striyaḥ.

Also *Bṛd.* from 1.121-*yaścaindro madhyamasthānogaṇaḥ so' yamataḥ paraḥ to* 2.6.

18. As to principal deities, see *Nir.* (7.5) *sūryo dyusthānaḥ;*

Also *Bṛhad.* (2.7): *yaḥ parastu sauryo dyusthānastaṁ nibodhat.*

For co-deities, see Nir. (7.10) -

athaitānyāditya bhaktīni asau lokaḥ, tṛtīyasavanaṁ, varṣā, jagati, saptadaśa stomaḥ, vairupaṁ sāma, ye ca devagaṇāḥ samāmnātā uttame sthāne yāś' ca striyaḥ.

See also *Bṛhad.* 2.8 to 2.16.

19. *RV.* 1.197.1; *VS.* 8.4; 33.66; *TS.* 1.4.22.1.
20. *RV.* 3.32.12.
21. *VS.* 28.44. TB. 2.6.20.1.
22. *RV.* 1.26.6; SV 2.968.
23. *RV.* 8.27.13; VS. 33.91.
24. 8.23.22.
25. *Nirukta*
26. *RV.* 1.1.1.
27. *RV.* 5.11.2; SV. 909; TS. 4.4.4.3.
28. *RV.* 2.2.1.
29. *Kau.* Br. 3.6; 5.5; Go. Br. 2.1.23; JB. 1.93; Tāṇ. Br. 6.1.6.
30. *JB.* 3.300.
31. *KS.* 4.16; AB. 1.4.8.
32. *TA.* 10.35. 73.33. Ś. Br. 1.6.3.39.
33. *Ś. Br.* 8.1.2.10; 9.3.2.7.
34. 1.6.5.

35. *RV.* 10.8.6.

36. *VS.* 6.16.

37. *agniṁ ca havyavāhanam.* *RV.* 2.41.19.

38. *agnaye kavyavāhanāya manthaḥ. KS.* 9.6. *agnaye kavyavāhanāya svadhā namaḥ. AV.* 18.4.71; TB. 1.3.10.3; ĀŚv. Śr. Su. 2.6.12; Āp. Śr. Su. 18.4; MS. 1.1.2.18.

39. *agniṁ dūtaṁ prati yadabravīt naḥ. RV.* 1.161.3. *agniṁ dūtaṁ vṛṇimahe. RV.* 1.12.1; AV. 20.101.1; SV. 1.3. *agniṁ dūtaṁ puro dadhe. RV.* 8.44.3. *agnireva devānām dūta āsa. Ś. Br.* 3.5.1.21.

40. *AB.* 7.9.8.

41. *RV.* 8.102.7; *SV.* 21; 146.

42. *RV.* 10.20.8.

43. *SS.* 1.14.6.8.

44. *Ś. Br.* 2.2.3.24; 5.2.31.

45. *AG.* 1.21.1; ŚG. 2.10.3; SMB 1.6.32.

46. *TS.* 6.3.7.1; MS. 1.4.11; 1.59.9; Ś. BR. 1.3.5.2; 2.5.2.19.

47. *RV.* 5.14.6.

48. *Kāṭh.* S. 35.3.

49. *TS.* 3.1.45.

50. *Manusmṛti,* 3.67.

51. *Kāṇ.* Ś. Br. 7.3.4.20.

52. *Kau.* Br. 12.4.

14

Rainmaking
A Historical Perspective

Rainmaking is not an idea of recent origin that has downed upon the scholars of the twentieth century only. Since long people have been attempting to induce rain by a variety of methods such as the rain dance, singing of rain songs, offering prayers, the sacrifice of men or animals lighting of fires, firing of cannon and the production of electric discharges by kites. But none of these methods could stand the test of time and proved to be worthless exercises and so by and by receded to the background and went out of vogue. All of those methods were based on the false belief and assumptions of the people and had no concrete or scientific basis behind them.

Modern Method of Rainmaking

The methods of modern rainmaking are, however, said to be based on a knowledge of physical processes of rain-formation. Modern rainmaking experiments are based on three main assumptions,

1. That either the presence of ice crystals in a supercooled cloud is necessary to release snow and rain by the Wagener-Bergeron process, or the presence of comparatively large water droplets is necessary to initiate the coalescence process.

2. That some clouds precipitate inefficiently or not all because these agents are naturally deficient.

3. That deficiency can be remedied by seeding clouds artificially with either solid carbon dioxide (dry ice) or silver iodide to produce ice crystals, or by introducing water

droplets or large hygroscopic nuclei.[1]

Actually, the history of modern rainmaking marked its first beginning in 1931 when Varaat of Holland dropped dry ice, among other things, into super-cooled clouds and produced slight amounts of rain. The possibility of producing rain from supercooled clouds by the introduction of artificial nuclei was also foreseen by the German cloud physicist Findeisen in 1938, but it was not until 1946 when a satisfactory method of supplying nuclei in the required quantity was discovered. The first time, Vincent J. Schaefer, an American meteorologist and science consultant succeeded in putting 6 pounds of crushed dry ice into a supercooled altocumulus cloud near Greylock Mt. in western Massachusetts on Nov. 13, 1946. In 1947, Vennegut discovered the effectiveness of silver-iodide as a cloud seeding material. This was followed the next year by langmuir's demonstrations that plain water could sometimes trigger the precipitation process in warm clouds by developing a sort of chain reaction in droplet growth. Since then many efforts have been made by meteorologists to find a cloud-seeding material better than silver iodide or dry ice, they are still only materials currently in use for large-scale operations directed towards modifying supercooled clouds.

Though, during this long span of time, a change in techniques of introducing freezing nuclei into clouds has been observed. First of all, freezing nuclei like dry ice or silver iodide were delivered only by airplane into the clouds. Later a search for better freezing nuclei than dry ice or silver iodide led to the disclosure that smoke of silver-iodide also acts as still a better freezing nuclei at temperatures as high as -6^0C. Then it was considered not to be necessary to apply the expensive technique of delivering silver-iodide by airplane as usual, but it could be introduced in the form of smoke from the ground itself with the help of ground generators or surface burners. This idea was conceived from the Indian tradition of performing *Yajñas.* Actually, these ground generators are the modern representatives of the ancient *Yajña kuṇḍas* or fire altars.

The technique of seeding clouds with the help of surface burners is very cheap compared to that of airplane or firing artillery shells, a very old idea revived by Russian scientists. The technique of *Yajña-kuṇḍas* or surface burners has only one drawback that is to rely on the air currents to carry the smoke up into the atmosphere. In this method, with no control over the subsequent transport of the smoke, it is not possible to make a reliable estimate of the smoke reaching into upper spheres. To overcome this problem, the smoke can be dispersed from burners in aircraft. This idea was also being practised in ancient India, as has been revealed in the *Bṛhadvimānaśāstra,* an ancient Indian treatise on the science of aeronautics. Airplanes were equipped with the *Dhūmodgama* or *Dhūmaprasāraṇa* instruments to disperse the smoke in the atmosphere so that the ill-designed bids of the enemy might be foiled, e.g. one verse of the *Bṛhadvimānaśāstra* (15.65) reveals how the airplane was used to be protected from enemies' attack by dispersing poisonous smoke from the *Apasmāradhūmaprasāraṇa Yantra* fitted in the plane.

svayāna rakṣaṇārthāya parayānair yathāvidhi
apasmāra dhūmaprasāraṇa yantraṁ pracakṣate.

In spite of the best thorough-going efforts of meteorologists in the field of rainmaking, it can be concluded that the modern science and technology of rainmaking is still in its infancy. It is confined only up to cloud modification. It has not so far been able to go beyond this. It doesn't see any prospects or scope for cloud-formation. Moreover, the results of trials carried out in the direction of cloud seeding have not been so encouraging and impressive. To sum up, it can be said that modern scientists have just made a beginning in the direction of weather modification and still they are to go a long way up to give a concrete and scientific shape to their immature start. Perhaps, that is why, they have not been able even to speculate to modify the direction of winds, to form clouds, to bring seasonal changes and to ward off the falling rain. Thus even after long consistent efforts of study and research into the field of weather modification, the achievements of the modern science of weather modification may be calculated as nil.

Rainmaking with the help of Yajña

In addition to the various efforts made in the direction of rainmaking as laid down above, one more attempt has been made since the time immemorial and that is with the help of *Yajña.* This technique has a long-standing history and has reached us through an age-old long tradition. That tradition is still alive in the historical records of our *Saṁhitās, Brāhmaṇas. Āraṇyakas, Upaniṣads, Sūtras,* Epics and *Pūraṇas* and other medieval and modern literature pertaining to the *Vedas, Upavedas* or *Vedāṅgas.* First of all, in the history of humankind, the idea of inducing rain with the help of *Yajña* was conceived by a Vedic seer, who in the very explicit terms proclaimed as *vṛṣṭiśca me yajñena kalpatām.*[2] 'Get me rains with the help of *Yajña.*' Perhaps, keeping in view the authenticity of this technique, another Ṛṣi proclaimed '*nikāme nikāme naḥ parjanyo varṣatu'.*[3] 'Let the cloud precipitate as and when we desire.' This science of rainmaking didn't end with these proclamations of the Vedic Ṛṣis, but in course, rainmaking agents were also discovered and it was made known that *Mitra* and *Varuṇa* were the rainmaking agents.

mitrāvaruṇau tvā vṛṣṭyāvatām.[4]

'May the *Mitra* and *Varuṇa* bring rain for you.'

'mitrāvaruṇau vṛṣṭyādhipati tau māvatām.[5]

'May the rainmaking agents, *Mitra* and *Varuṇa,* protect you.'

The other Ṛṣis made further advances in the area and could come to the conclusion that the co-ordination of both the elements is necessary for inducing rain and the proposed coordination could easily be effected with the help of *Yajña.*

yajñā no mitrā varuṇā yajñā devam ṛtaṁ bṛhat.[6]

'*Mitra* and *Varuṇa,'* the main agents of rain, should be

coordinated with the help of *Yajña* for the precipitation of rain

Thus the *Yajña* became the part and parcel, in fact, the basis of the Vedic life. Perhaps, that is why, the Vedic Ṛṣi didn't hesitate in passing the direction to every human being to perform *Yajña* season-wise daily, monthly and yearly to stimulate seasonal deities so that they may yield their favour for them as per their desires.

ṛtūn yajña ṛtupatinārtvānut hāyanān
samāḥ saṁvatsarān māsān bhūtasya pataye yajña.[7]

Following the notion of the Vedic Ṛṣi, the author of the *Brāhmaṇas* proclaimed *Yajña* as the noblest act.

yajño vai śreṣṭhatamaṁ karma.[8]

They also elucidated the process and role of *Yajña* in stimulating precipitation as:

agnervaidhūmo jāyate. dhūmād abhram. abhrād vṛṣṭ iḥ.[9]

'*Yajñīya* fire helps in smoke development, the smoke helps in cloud formation and clouds yield rain.'

The idea of rainmaking with the help of *Yajña,* which found its genesis in the *Saṁhitās* was developed in the *Brāhmaṇas* and *Āraṇykyas* and sought its culmination further in the *Upaniṣadika* literature. The *Upaniṣadika* Ṛṣi at one place doesn't want to miss the opportunity of disclosing the technique as to how to induce an existing cloud to rain. He speaks thus:

parjanyo vāgnir gotam tasya saṁvatsara eva samidabhrāṇi dhūmo vidyudarcciraśanir aṅgārā hrādunayo visphuliṅgās-tasminnetasminn agnau devāḥ somaṁ rājānaṁ juhvati tasya āhutyai vṛṣṭi samabhavati.[10]

'That is to say briefly, an existing cloud may be stimulated to precipitate by the *Soma āhuti* released into the *Yajña.*'

The same idea was further inherited by the Epics and

Purāṇas. Lord Krishna in his message to the dejected and disappointed Arjuna while relating the overall dominance of *Yajña,* describes the fact that the *Yajña* is helpful in inducing rain. The verses go like this:

annād bhavanti bhūtāni
parjanyād anna saṁbhavaḥ.
yajñād bhavati parjanyo
yajñaḥ karma samudbhavaḥ.[11]

karma brahmodbhavaṁ viddhi,
brahmākṣara samudbhavam
tasmāt sarvagataṁ brahma nityam
yajñepratiṣṭhitam.[12]

In addition to this, the *Bṛhadvimānaśāstra* an ancient treatise on the science of aeronautics, treats this subject in a more technical and mechanised way. It gives detailed break-up as to how to develop various types of scientific instruments to save the airplane from all those natural calamities which are likely to cause damage to it and its crew. For instance, it proposes on the basis of *Yantrasarvasva* to develop a *Śiraḥkilaka* instrument to dispel the effect of lightning.

yadapāyo vimānasya bhavedaśanipātataḥ
tadapāya nivṛttya-rthaṁ śiraḥkilaka yantrakam.[13]

Further, it lays down the techniques as to how to develop, among several other instruments, the *Varṣ opasaṁhāra, Tryāsyavātanirsana* and *Ātapopasaṁhāra* instruments to neutralise the effects of rain, winds and heat respectively, which are likely to affect the aeroplane. e. g.

uktāni hi yantrasarvasve yantratrayaṁ yathā vidhi.[14]
sarveṣāṁ sukhabodhāya tānyevatra pracakṣate

tryāsyavāta nirasana yantraṁ tadvān manoharam.[15]
sūryātaposaṁhāra yantraṁ caiva tataḥparam

ativarṣopasaṁhāra yantraṁ ceti tridhā smṛtam.[16]

To protect the airplane from the opposition of strong air,

Vimānastambhan instrument was stated to have been installed in the airplane, e. g.

> *vātapravāhasaṁasarga parihārāya kevalam vimānastambhanayantraṁ yathāmati nirupyate.*[17]

Dayananda's Revelation

In modern times, the age-old idea of rainmaking which fizzled out with the passage of time was revived by Swami Dayananda Saraswati, a great embodiment of Vedic life and thought. He made revolutionary observations on the various burning issues concerning political, social, economic, religious, and linguistic aspects of life and on the interpretation of ancient Indian language and literature. Among an immense number of things, he observed on the basis of the evidence from the Vedas that the rain could be induced with the help of *Yajña* as and when we desire. During the course of relating various purposes of *Yajña,* he gives a detailed description of the process of inducing rain with the help of *Yajña.* He bases his ideas on the evidence of *Śatapatha Brāhmaṇa* 5.3.5.17. According to him,

> *agneḥ sakāśād dhūmavāṣpau jāyete. yadā yamagnirvṛkṣauṣadhivanaspati jalādipadārthān praviśya tān saṁhatān vibhidya tebhyo rasaṁ ca pṛthak karoti, punaṣte laghutvamāpannā vāyvādhār-eṇoparyyā kāśaṁ gacchanti.tatra yāvān jalarasā-ṁśastāvato vāṣpasaṁjñāsti. Yaśca niḥsneho bhāgaḥ sa pṛthivyaṁśo'sti. ata evobhaya bhāgayukto dhūma ityupacaryate. Punar dhūmagaman-ānantaramākāśe jalasaṁcayo bhavati. tasmādabharaṁ ghanā jāyante. tebhyo vāyudalebhyo vṛṣṭirjāyte.*[18]

'That is, due to the treatment with fire, carbon and vapour are generated. When the fire enters the trees, herbs, plants and water, etc. it decomposes them and separates their juicy substance from them, which on account of its lightness, is accelerated upward towards the sky in conjunction with air. In this process, the juicy substance which is an aqueous form is known as vapour and what remains dry is known as carbon which is the part of the earth. So both forms earthen and aqueous in the combined state are known as smoke. After the smoke reaches the sky, water vapours accumulate there. These vapours give birth to clouds and clouds yield rain.'

Ram Narain Arya as a Rainmaker

Following Dayananda's revolutionary observations, the idea of rainmaking with the help of *Yajña* gained a momentum and found new support at the hands of *Āryasamājī* scholars. A few of them even came forward who endeavoured to induce rain or to prevent it, but their experiments couldn't yield fruitful results.[19] Here in this regard, a particular mention may be made of Ram Narain Arya, M.A. who can be called as a perfect rainmaker of the twentieth century. After spending 35 long years of study, research and experimentation, he has come to a startling conclusion what he calls the Vedic way to beat the nature. Not only has he been able to induce rain successfully, but he has also been forming clouds, changing the flow of air in terms of direction and speed, stopping the falling rain and even modifying into the weather conditions, from dry to wet, hot to cold and vice-versa for the last 36 years. The author of the present lines has been a close associate and witness to most of the rainmaker's experiments carried out by him from time to time at various places in India. Having been confirmed strongly of the authenticity and success of the rainmaker's experiments, this author thought it viable to bring the rainmaker's research into the limelight so that not only the people of India but the people of the whole world may be benefitted by the experiences and research of the rainmaker. Actually, the rainmaker has preserved his ideas and experiences, with regard to his experiments on weather modification like rainmaking, prevention of rain, modification into the direction and speed of airflow, modification into the other weather conditions, prevention of pollution and diseases in his daily diaries written by him from time to time. This author could get the privilege to make good use of the subject matter enshrined in the daily diaries (DD) of the rainmaker. At the places of apprehension and doubts, he made serious discussions with the rainmaker to clarify the same. To proceed further, it is necessary to render hereunder the brief life-sketch and work of the rainmaker.

A Brief Life-sketch of the Rainmaker

Sh. Ram Narain Arya born on *Amāvaśyā* (the new moon day) of the *Srāvaṇa* month in 1993 of the *Vikram* era in the village of Farmana, District Sonepat, Haryana. His mother breathed her last when he was in his infancy. Under the circumstances, he along with his two brothers and one sister was brought up by his father. His father, Sh. Ratti Ram Arya was fond of physical exercises and wrestling. A devout and religious fellow, he was a devotee of the Veda. Ram Narain was greatly influenced by his father and inherited his qualities of doing *Yaga* and physical exercises regularly. He completed his schooling from the village high school. Afterwards, he passed out the examination of I.G.D. Bombay from Bareli and was appointed in 1955 as a Drawing teacher in the Govt. Higher Secondary School, Bahujholari (District Jhajjar of Haryana). Late in the forties, he completed his B.A. and did his M.A. in Political Science and Sanskrit (Veda). From childhood, he was very intelligent, curious and studious one. He had the chance to study the literature of Kabeer, Raheem, Tulsidas, Raidas and Guru Nanak. His wife, Prem Vati Prabhakar, who is also a religious lady, used to read him the stories of *Rāmāyaṇa*, *Mahābhārata* and *Gitā.* All this added to his detachment from the worldly allurements. *Gṛhastha* as he is, he is leading a life of an ascetic. In 1955, he came in contact with Bhagvandeva Acharya (Swami Omananda), Acharya Baladeva and Brahmachari Indradeva (Swami Indravesh) at Gurukula Jhajjar. This new acquaintance promulgated him to study *Vedas.* He then moved to the study of the *Vedas, Upavedas* and *Vedāṅgas.* Influenced by Swami Dayananda's works such as *Satyārhta Prakāsh* and *Saṁskāravidhi,* he started performing *Yajñas* daily at the time of sunset and sunrise. During his studies of the *Vedas* and Dayananda's works, he came across such references as could give him an idea of rainmaking with the help of *Yajña.* He also made an in-depth study of the works of Āryabhaṭṭa, Varāmihira and Bhāskarācārya which added to his knowledge of geology and astronomy. He had a chance to make an intensive study of the *Bṛhadvimānaśāstra,* an ancient Indian treatise on the science of aeronautics composed by Maharishi Bharadvāja. This work closeted him with the different aspects of ancient Indian science related to weather modification. Thus having taken cues of rainmaking with the help of *Yajña* from the Vedas, Upavedas, Vedāṅgas and Dayananda's works and having acquired the knowledge of the properties of the matter as described therein, he all set to perform *Yajñas* in a particular direction. It was to modify weather, modify airflow in terms of direction and speed, to induce rain and to prevent the same, to remove famines, to prevent deluges, diseases and to beat the pollution.

The experimental *Yajñas* performed strictly in conformity or in tune with the Vedic principles and surprisingly enough everything was witnessed taking place actually which was considered sometimes ago a fallacy or mythology. He calls this method of controlling natural powers with the help of *Yajña* as a *Vedic way to beat the nature.* Not only did he do experiments with nature outside, but he also practised *Yoga* to attain self-accomplishment and applied it to detect the deficiencies or inconsistencies in the matter of natural course. According to him, *Yajña* and *Yoga* should go side by side. By way of *Yoga,* one can feel the nerve of nature and detect its deficiencies or inconsistencies like an able and dexterous physician does with a patient and by way of *Yajña,* medication of nature can be done to remedy all its deficiencies and inconsistencies. His dedication to this cause was so great that all the institutes he served were converted by him into laboratories for his experiments on nature. He served at the following institutions. G.H.S.S. Bahujholari (1955-66); G.M.S. Ruraki (1967-68); G.M.S. Guhna (1-4-1968 to 24-9-1968); G.H.S. Khanpur Kalan (25-9-68 to 7-10-68); G.H.S.

Mundalana (11-10-1968 to 18-7-1973); G.M.S. Rithal (19-7-1973 to 31-7-1974); G.G.H.S. Bhainswal Kalan (1-8-1974 to 18-5-1976); G.M.S. Katwal (19-5-1976 to 1-8-1978); G.H.S. Bhainswal Kalan (2-8-1978 to 4-7-1979); G.M.S. Jasrana (5-7-1979 to 7-9-1989); G.H.S. Anwali (8-8-89 to 8-12-90); G.H.S. Bichpari (9-12-90 to 21-7-91). All these schools have been the main centre of his experiments.

Having been confirmed of the validity and authenticity of the Vedic principles, he became a staunch exponent of the Vedic life and thought. He made it his life mission to propagate Vedic teachings and thought among the general masses, particularly youths and for that matter, he used to deliver prayer time discourses before students on the Vedas and their ancillary sciences during his teaching days. He, thus, made them known about the glorious past of India and ancient Indian advancement in the field of philosophy, sociology, polity, theology, science and technology. He voluntarily retired from the Govt. service in 1991 in order to speed up his activities for the furtherance of this noble cause. After retirement, he speeded up his mission of spreading scientific knowledge of the Vedas. He toured extensively and delivered more than 1000 public and popular lectures in different educational institutions and Jails to enlighten the students and jailed persons of the rich scientific heritage of ancient India and exhorted them to go for *Yoga* and *Yajña* for better health and healthy environment in society. He warns the younger generation not to fall prey to the cultural pollution that is posing a great danger to the humanity at large. He himself leads a very simple and austere life. His whole life has been a life of relinquishment and penances. Through *Yoga* and penances, he has so regulated his life that he spends the chilly winters away in simple summer wears. He takes simple *sātvika* meals often saltless, chilly-less and sugarless. According to him, prior to the regulation of nature, regulation of the self is a must.

An Account of the efforts made by the Rainmaker

The rainmaker, Ram Narain Arya, has spent 55 long years of his consistent integrated study and research into the Vedic texts. His studies and researches have led to the conclusion that the Vedic way of *Yajña* to beat the nature in various spheres is quite reliable, concrete and scientific one. Within this span of time, he has conducted over 300 experiments (major and minor ones together) to induce rain and abandon the same whenever and wherever necessary. A brief account of his experiments and newspaper reports on the subject may shed an ample good light on the keenness of his study, depth of his research and his zeal for the promotion of this science of the Vedas for the well-being of humankind.

Experiments

It was the beginning of 1955 when Ram Narain Arya started his clandestine experiments of rain and anti-rain at a personal level. By the time of 1966, he was confirmed and had confidence that rains could be induced and abandoned with the help of a Vedic scientific *Yajña* with the special type of oblatory material particularly used for rainmaking and anti-rain. Below is given a chronological account of experiments performed by him from time to time.

1. 7-9-1972: (Rain-*yajña*): By the end of 1972, it was thought justifiable to bring this research into the limelight, so that the general public might also be benefitted by it and with that view, the first open Rain-*yajña* in public was performed in the village Farmana, Sonepat, Haryana from 7-9-1972 to 10-9-1972. The rain started pouring down on 10-9-72 and continued till 14-9-1972 (vide DD[1]-1, P.1).
2. 17-9-1972: (Rain-*yajña*): The 2nd experiment was demonstrated in the G.H.S. Mudalana, Sonepat, Haryana when the sky was clouded and east wind was

---

[1] DD here means Daily diaries written by Rainmaker from time to time.

blowing day and night. It took only 24 hours to induce the rainfall at the place of the experiment at the cost of Rs. 100/- only. (vide DD-1, P.1).

3. 2-10-1972: (Change of the wind, cloud formation and rain induction): A constant flow of west-wind was modified to eastward and afterwards clouds were formed within 24 hours that led to the raining. The *Yajña* was performed from 2-10-1972 to 8-10-1972 (vide DD-1, PP. 4-7).
4. 30-2-1992: Modification of Weather and rain formation was demonstrated at Jhajjar Road, Rohtak (vide DD-1, PP.4-7).
5. 15-1-1973: Anti-rain experiment was performed at Rohtak as well as Farmana, Distt. Sonepat, Haryana from 15-1-73 to 17-1-1973 and the rains were completely abandoned. (vide DD-1, P. 39-40).
6. 14-7-1973: Rain-*yajña* was performed from 14-7-1973 to 21-7-1973 in V.P.O. Siwani, Distt. Bhiwani, Haryana at the invitation of the people. It started raining from 15-7-1973 itself and whole of the District experienced over *kyārī*-full (field-bed full) of rain. (vide DD-1, PP. 45-46).
7. 20-5-1974: Rain-experiment was carried out before the teachers in G.M.S. Rithal, Distt. Sonepat, Haryana where rainfall was made to occur within 24 hours when the east wind was sweeping and the sky was clouded. (vide DD-1, PP. 47-48).
8. 30-6-1974: Rain-y*ajña* was performed in Nizampur Majra, Distt. Sonepat, Haryana from 30-6-1974 to 7-7-1974. Clouding started from 1-7-74 which ultimately led to raining from 3-7-1974, 4-7-1974 and so on up to 16-7-1974 with some intervals of days. (vide DD-1, P. 48).
9. 16-1-1974: Rain experiment was conducted in Farmana, Distt. Sonepat, Haryana from 16-9-1974 to 24-9-1974. It started raining from 18-9-1974 and continued on 21-9-1974, 24-91974 and 25-9-1974. (vide DD-1, PP. 51-

53).

10. 4-8-1975: Anti-rain experiment was conducted in Farmana, Distt. Sonepat, Haryana from 4-8-1975 to 10-8-1975 which helped prevent rain although the period of 7 days of the experiment. (vide DD-1, PP. 55-57).
11. 11-8-1975: Rain-*yajña* was started in Farmana, Distt. Sonepat, Haryana from 11-8-1975 to 16-8-1975, consequent upon which it started drizzling on 12-8-1975 and since the area experienced heavy rainfall, it was decided that an anit-rain *Yajña* be performed. (vide DD-1, PP. 57-58)
12. 29-8-1975: Rain-*yajña* was performed in Farmana, Distt. Sonepat, Haryana from 29-8-1975 to 7-9-1975. As a result, good amount of rain was recorded on 29-8-1975, 6-9-1975 and 7-9-1975. (vide DD-1, PP. 75-76).
13. 1-9-1975: Rain-*yajña* was conducted again in Farmana, Distt. Sonepat, Haryana from 1-9-1975 to 6-9-1975 that led to sufficient rain-fall till 7-9-1975. (vide DD-1, PP. 77-78).
14. 7-9-1975: Anti-rain experiment was done in Farmana, Distt. Sonepat, Haryana from 7-9-1975 to 18-9-1975 (which caused the rain stopped completely for the required period. (vide DD-1, PP. 78-79).
15. 27-1-1976: Rain-experiment was carried out in Govt. Girls' High School Bhainswal Kalan, Distt. Sonepat, Haryana from 27-1-1976 to 1-2-1976 which resulted in the good amount of rain-fall on 28-1-1976 and 1-2-1976 (vide DD-1, P.88).
16. 18-2-1976: Anti-rain demonstration was made in G.G.H.S. Bhainswal Kalan, Distt. Sonepat, Haryana itself from 18-2-1976 to 19-2-1976 when the east wind was in sway and the sky was overcast with clouds. The experiment paralysed the downpour of rain. (vide DD-1, P. 89).
17. 22-2-1976: Anti-rain demonstration was also given in

Farmana, Distt. Sonepat, Haryana on 22-2-1976 which was also a success. (vide DD-1, P. 90).

18. 24-2-1976: *Yajña* for the change of weather was performed in Farmana, Distt. Sonepat, Haryana from 24-2-1976 to 28-3-1976 which changed the rainy weather into the dry one. (vide DD-1, P. 90).
19. 3-8-1976: Anti-rain demonstration was given on 3-8-1976 in G.M.S. Katwal, Distt. Sonepat, Haryana which brought the falling rain to an end for the day. (vide DD-1, P. 93).
20. 4-8-1976: Anti-rain experiment was conducted again the following day in G.M.S. Katwal, Distt. Sonepat, Haryana itself. It was raining heavily and east wind was blowing. As soon as the *Yajña* was started, dark clouds turned white and severe rain showers were reduced to drizzling which ultimately came to an end. (vide DD-1, P. 94).
21. 22-8-1976: Rain-experiment was performed in Farmana, Distt. Sonepat, Haryana on 22-8-1976 and the success was achieved in inducing rain the next day, i.e. 23-8-1976. (vide DD-1, PP. 94-95).
22. 24-8-1976: Anti-rain *Yajña* was conducted in Farmana, Distt. Sonepat, Haryana from 24-8-1976 to 5-9-1976 which helped the rain-fall stop for the desired period of the experiment. (vide DD-1, PP. 95-96).
23. March 1977: (Change of Weather). From March 1977 to June 1977 such experiments were carried out as could maintain coldness during the period of scorching hot with a little bit of drizzling at intervals. (vide DD-1, PP. 100-102).
24. 15-7-1977: Anti-rain experiment was performed in Farmana, Distt. Sonepat, Haryana from 15-7-1977 to 22-7-1977 in order to stop unwanted heavy rain, which was likely to cause heavy damage to the houses of the villagers and crops. (As to details see DD-1, PP. 122-123).

25. 23-7-1977: Rain-*yajña* was performed from 23-7-1977 to 30-7-1977 to put an end to the scarcity of rain at Farmana, Distt. Sonepat, Haryana. (vide DD-1, PP. 133-135).
26. 8-8-1977: Anti-rain experiment was carried out at Farmana for the fear of damage which was likely to be caused by a heavy downpour of rain. (For details see DD-2, PP. 9-10).
27. 19-8-1977: Anti-rain experiment was again conducted at Farmana when all exits of the village were blocked by rainy waters caused by uninterrupted heavy rains. (vide DD-1, PP. 16-19).
28. 21-8-1977: Anti-rain experiment was performed at Farmana when it was raining cats and dogs. (vide DD-2, PP. 22-26).
29. 8-3-1978: Anti-rain experiments were carried out from 8-3-1978 to 11-3-1978 at Farmana in order to control floods caused by excessive rains and to save crops from irreparable loss. (vide DD-2, P. 67).
30. 2-2-1979: Anti-rain *Yajña* was performed at Gurukula Singhpura, Distt. Rohtak, Haryana for three days, i.e. from 2-2-1979 to 4-2-1979 in order to save the annual function of the Gurukula from disruption by heavy rains. Wind was modified from eastward to westward and the rain was completely brought to an end throughout the function which lasted for three days. (vide DD-2, p. 80).
31. 28-3-1979: The *Yajña* for prevention of rains as well as diseases was performed at Farmana from 28-3-1979 to 30-3-1979 in order to check the spread of diseases in the village feared to be caused by excessive rain and it was a success. (As to details see DD-2, PP. 87-88).
32. 13-4-1979: Rain-*yajña* was performed on 13-4-1979 at Farmana to test the worth of some herbal substance in inducing rain. (vide DD-2, P. 91).
33. 14-4-1979: Anti-rain experiment was carried out on

14-4-1979 at Farmana to stop the rain-induced by the preceding rain *Yajña.* (vide DD-2, P.91).

34. 22-8-1979: Rain-experiment was performed from 22-8-1979 to 31-8-1979 at Farmana, consequent upon which the heavy rain took place. (vide DD-2, PP. 111-115).
35. 1-9-1979: Anti-rain experiment was necessitated to check the heavy downpour of rain caused by the earlier rain experiment performed from 1-9-1979 to 7-9-1979 at Farmana itself. (vide DD-2, PP. 111-115).
36. 8-9-1979: Rain-*yajña* was again performed from 8-9-1979 to 15-9-1979. The rain started pouring down from 10-9-1979 and the heavy rain was reported on 11-9-1979 and 14-9-1979 which continued up to 19-9-1979. (For more details, see DD-2, PP. 111-115).
37. 25-8-1979: Anti-rain experiment was performed from 25-8-1979 to 28-8-1979 at the villages of Farmana, Nizamppur-Majra and Mungan to stop the unwanted rain. (vide DD-2, PP. 101-105).
38. 30-8-1979: Rain-experiment was performed with the assistance of the villagers at Farmana from 30-8-1979 to 11-9-1979. Lightning started flashing on the starting day itself and it drizzled throughout the intervening night. The rain was reported on 3-9-1979 and 5-9-1979. (vide DD-2, PP. 109-11).
39. 17-11-1979: *Yajña* for cloud and rain formation was performed on 17-11-1979 at Nandana, Distt. Rohtak, Haryana. (vide DD-2, P. 229).
40. 16-2-1980: Rain-*yajña* was performed on 16-2-1980 at Farmana on the occasion of sun-eclipse which led to the raining (from low to moderate) althrough the northern region. (vide DD-2, P. 259).
41. 23-7-1980: Anti-rain experiment was conducted from 23-7-1980 to 27-7-1980 at Farmana which completely paralysed rain for the stipulated period. (vide DD-2, PP. 276-277).

42. 11-8-1980: Anti-rain *yajña* was performed from 11-8-1980 to 28-8-1980 at Farmana to prevent the occurrence of the rainfall. (vide DD-2, PP. 251-284).
43. 1-9-1980: Rain-experiment was carried out from 1-9-1980 to 4-9-1980 at Farmana. Delhi, Panjab and Haryana experienced a good amount of rain up to 8-9-1980. (vide DD-2, PP. 284-286).
44. 8-9-1980: Anti-rain experiment was conducted from 8-9-1980 to 15-9-1980 at G.M.S. Jasrana, Distt. Sonepat, Haryana to stop the occurrence of rain-fall induced by the preceding experiment for rain. It was also a success. (vide DD-3, PP. 5.21).
45. 16-9-1980: Rain-*yajña* was performed from 16-9-1980 to 30-9-1980 at G.M.S. Jasrana, Distt. Sonepat, Haryana to dispel the effect of Anti-rain experiment. From 17-9-1980, it started raining and continued through 19-9-1989, 28-9-1980 till 30-9-1980. (vide DD-2. 21-51).
46. 8-10-1980: Rain-experiment was performed from 8-10-1980 to 14-10-1980 at Farmana. It started raining from 14-10-1980 and continued up to 27-10-1980 with varied-low, moderated and heavy spells. (vide DD-3, PP. 55).
47. 13-12-1980: Anti-rain experiment was carried out from 13-12-1980 to 15-1-1981 at Farmana, which banned the downpour of rain. (vide DD-3, P. 66).
48. 16-1-1981: Rain-*yajña* was performed from 16-1-1981 to 31-1-1981 at Farmana. It rained on 24-1-1981 and 30-1-1981. (vide DD-3, P. 66).
49. 1-2-1981: Anti-rain experiment was conducted from 1-2-1981 to 15-2-1981 at Farmana. The rain was completely stopped except a few spells on 5-2-1981 and 10-2-1981. The sky remained clear. (vide DD-3, PP. 67-68).
50. 16-2-1981: Rain-experiment was conducted from 16-2-1981 to 28-2-1981 which caused rain in Delhi, Haryana

and some parts of U.P. and Rajasthan. (vide DD-3, PP. 67-71).

51. 11-3-1981: Anti-rain *Yajña* was performed from 11-3-1981 to 31-3-1981 in Farmana and G.M.S. Jasrana, Distt. Sonepat, Haryana which brought the rain to an end. (vide DD-3, PP. 148- 159).
52. April 1981: Anti-rain experiment was extended up to April 1981 keeping in view the excessive rainfall. (vide. DD-3, PP. 160-167).
53. May 1981: (Change of weather and rain formation)-In order to change the hot and dry weather of May into cold and rainy one, the *Yajña* was performed for the period of May in Farmana and the goal was achieved. (vide DD-3, PP. 168-175).
54. June 1981: The above experiment was extended further up to the next month, i.e. the June of 1981. By the and of June, heavy spells of rain were experienced, hence it was resolved that an Anti-rain experiment would be conducted. (vide DD-3, PP. 177-192).
55. 1-7-1981: Anti-rain experiment was started from 1-7-1981 in order to contain the floods which were feared to be caused by excessive rain induced by the two months long previous experiment and to bring down the spells of rain from heavy to low. It was so done as to help the millet and barley grow better. (vide DD-3, PP. 192-202).
56. 12-9-1981: Rain-*yajña* was performed from 12-9-1981 to 21-9-1981 at Jasrana, Farmana, Majra and Asan. (Rohtak) with the assistance of villagers. The rain started pouring down on 14-9-1981 and continued up to 22-9-1981. (vide DD-3, 204-218).
57. 2-10-1981: Anti-rain experiment was carried out from 2-10-1981 to 25-10-1981 at G.M.S. Jasrana and Farmana in order to check the untimely rainfall. (vide DD-3, PP. 225-236).
58. 26-12-1981: Rain-experiment was done from 26-12-

1981 to 2-1-1982 at Farmana, Majra and Guhna. The rain started falling heavily w.e.f. the night of 3-1-1982. (vide DD-3, PP. 258-271).

59. 6-3-1982: Anti-rain *Yajña* was performed to save the crops from the excessive rain at the proposal of the teachers. It was performed for three days, i.e. on 6-3-1982, 7-3-1982 and 8-3-1982 at G.M.S. Jasrana, Distt. Sonepat, Haryana. (vide DD-3, P. 286).
60. 13-6-1982: Rain-experiment was conducted from 1-6-1982 to 27-6-1982 at Farmana which succeeded in inducing heavy rain on 27-6-1982. (vide DD-4, PP. 7-10).
61. 5-8-1982: Rain-experiment was performed from 5-8-1982 to 18.8.1982 at Farmana, and as a consequence, heavy rainfall occurred at the place of experiment and the surrounding area from 17-8-1982 to 19-8-1982. (vide DD-4, PP. 26-32).
62. 29-10-1982: Rain-*yajña* was performed from 29-10-1982 to 31-10-1982 at Farmana which induced the low rain-fall on 30-10-1982 and 31-10-1982. (vide DD-4, P. 47).
63. 25-12-1982: Rain-experiment was conducted from 25-12-1982 to 1-1-1983 at Farmana, with the help of which the rain could be induced on 26-12-1982 and 28-12-1982. The rain was reported from all over north India. (vide DD-4, PP. 60-61).
64. 4-1-1983: Rain-*yajña* was performed from 4-1-1983 to 9-1-1983 at various places, *viz.* Farmana (on 4-1-1983), G.P.S. Ridhau (on 5-1-1983) and (6-1-1983), again Farmana (on 7-7-1983), G.H.S. Gorad (on 8-1-1983) and yet again at Farmana (on 9-1-1983). From low to heavy stormy rain was recorded on 9-1-1983 at all the places of experiments and the surrounding area. (vide DD-4, PP. 62-71).
65. 24-1-1983: Rain-experiments were performed on 24-1-1983 and 25-1-1983 at Ruraki, Distt. Rohtak, and Kasandi, Distt. Sonepat, Haryana respectively which

resulted in drizzling and low rain on 2-1-1983 and 27-1-1983 and moderate to heavy rain on 28-1-1983. (vide DD-4, PP. 76-80).

66. 16-2-1983: Rain-experiment was done from 16-2-1983 to 23-2-1983 at G.M.S. Jasrana, Distt. Sonepat, Haryana which succeeded in bringing rain up to the night of 23-2-1983. (vide DD-4, PP. 83-88).
67. 1-3-1983: Rain-demonstration was given on 1-3-1983 at Farmana, consequent upon which drizzling took place on 2-3-1983. (vide DD-4, PP. 83-88).
68. 10-5-1983: Rain and Anti-rain experiment were performed in the close succession from 10-5-1983 to 13-5-1983 at Farmana. The people were taken aback to see the induction of rain first and its prevention subsequently. (vide DD-4, PP. 138-143).
69. 18-15-1983: Anti-rain experiment was carried out from 18-5-1983 to 31-5-1983 at Farmana in order to facilitate people harvest their ripened crops and consequently there was no rain all through. (vide DD-4, PP. 154-164).
70. 5-6-1983: Rain-experiment was performed from 5-6-1983 to 12-6-1983 at Gurukala Indraprastha, Faridabad, Haryana. The rain started falling on 9-6-1983 and continued up to 13-6-1983 throughout Haryana. Swamy Shaktivesh, Incharge of the Gurukula also issued a certificate to the rainmaker on the grand success of the experiment. (vide DD-4, PP. 170-178).
71. 27-7-1983: Anti-rain experiment was conducted from 27-7-1983 to 7-8-1983 to put a check on heavy rain caused by monsoons. Till 4-8-1983 a total ban on rain-fall could be imposed, but on 5-8-1983 it experienced a few spells. Afterwards, from 6-8-1983 to 13-8-1983, there was no rain at all. (vide DD-4, PP. 186-202).
72. 14-2-1984: Rain-experiment was performed from 14-2-1984 to 20-2-1984 at G.M.S. Jasrana, Sonepat, Haryana. The sky started clouding from 15-2-1984 and the rain occurred on 18-2-1984 and 19-2-1984. (vide

DD-4, PP. 217-219).

73. 21-3-1984: Anti-rain experiment was done from 21-3-1984 to 29-5-1984 at Farmana. It was reported by the *Rājadharma,* Rohtak, Haryana. (vide DD-4, PP. 235-271).
74. 30-5-1984: Rain experiment was performed from 30-5-1984 to 8-6-1984 at the Yadav Seeds Farm, Baharampur-Fazalpur (Gurgaon) Haryana at the behest of Rao Virendra, Minister of Agriculture, Govt. of India. The sky started clouding from 1-6-1984 and the rain was reported from Haryana, Punjab and Delhi from 3-6-1984 onward which continued with intervals up to 14-6-1984, but the place of the experiment could receive a few rainy spells only for one day. (vide DD-4, PP. 273-334).
75. 8-8-1984: Rain-experiment was performed from 8-8-1984 to 13-8-1984 at Govt. High School Rabhra, Polangi and Ruraki, Haryana. The rain began to fall on 10-8-1984 and continued up to 15-8-1984. (vide DD-4, PP. 332-334).
76. 19-8-1984: Anti - rain *yajña* was also demonstrated on 19-8-1984 at Nimbri, Panipat, Haryana, which in consequence induced rain for the day of the experiment. (vide DD-4, PP. 332-334).
77. 27-8-1984: Anti-rain experiment was carried out on 27-8-1984 at Polangi, Distt. Rohtak, Haryana which paralysed the rain completely. (vide DD-4, PP. 332-334).
78. 8-11-1984: Rain-demonstration was given on 8-11-1984 at Ridhau, Distt. Sonepat, Haryana which induced drizzling in the intervening night. (vide DD-4, PP. 332-334).
79. 27-1-1985: Rain-experiment was performed from 27-1-1985 to 3-2-1985 at Asra -Kheri, Distt. Rohtak, Haryana. Clouds were formed within 24 hours of starting of the *Yajña.* It started raining all over Haryana from 28-1-1985. The place of experiment experienced

rain along with hailstorm on 3-2-1985. (vide DD-5, PP. 22-24).

80. 3-10-1985: Rain and Anti-rain experiment were conducted in close succession (i.e. from 3-10-1985 to 9-10-1985 Rain experiment and from 10-10-1985 to 24-10-1985, Anti-rain experiment) in Kastrubadham, Gujrat. Under the influence of Rain-experiment, it rained from 8 inches to 22 inches all over Gujrat. Anti - rain experiment stopped rain-fall to occur, clouds were visible all through day and night. (vide DD-5, PP. 47-67).
81. 14-2-1986: Anti-rain experiment was demonstrated from 14-2-1986 to 15-2-1986 at Asan, Distt. Rohtak, Haryana to check the heavy downpour of rain in order that the ritual of marriage could be performed uninterruptedly without any disruption whatsoever. (vide DD-5, P. 65).
82. 9-6-1986: Rain-experiment was performed from 9-6-1986 to 16-6-1986 at Hathi Khana Arya Samaj Mandir, Rajkot, Gujrat at the invitation of Hari Bhai Patel, the then Director of Agriculture, which induced the rain up to the required degree. This experiment was reported by the Gujarati dailies *Foolchhap* and *jai Hind* on 16-6-1986 and 21-9-1986 respectively.
83. 17-6-1986: Rain-*yajña* was performed from 17-6-1986 to 22-6-1986 in continuation of the previous on at *Upadeshaka Mahāvidyālaya,* Tankara (Maurvi State), Gujrat, the success of which was also reported by a Gujarati daily *loksatta* on 19-6-1986.
84. 4-8-1986: Rain-*yajña* was performed from 4-8-1986 to 10-8-1986 at Gorar, Distt. Rohtak, Haryana. As a result, lightning started flashing from 5-8-1986 and the rainy spells were witnessed at the place of experiment and the surrounding area on 11-8-1986 which continued till 13-8-1986 with varying degrees. (vide DD-5, PP. 76-79).
85. 12-8-1986: Rain-experiment was performed from 12-8-

1986 to 19-8-1986 at Nirena (Kachchabhuj), Gujrat, which was reported by a local Gujarati paper, Kachcha Mitra on 19-8-1986.

86. 23-7-1987: Rain-*yajña* was performed from 23-7-1987 to 28-7-1987 at Rani Talab Jind, Haryana. It started drizzling on 26-7-1987 and sufficient amount of rain was recorded by 28-7-1987. (vide DD-6. PP. 183-185).
87. 7-8-1987: Rain-experiment was done from 7-8-1987 to 16-7-1987 at *Prabhu Ashrita Arya Yoga Ashram,* Rohtak, Haryana as the arrange by Mahashaya Hardwarilal. The rain-fall was recorded on 13-8-1987, 16-8-1987 and 18-8-1987. (vide DD-6, PP. 183-185).
88. 17-8-1987: (Change of wind)-An experiment to maintain the east-wind was carried out at the invitation of Acharya Baldev at Gurukala Kalwa, Distt. Jind, Haryana from 18-6-1987 to 23-8-1987. It induced rain on 22 and 23-8-1987 which lasted up to 29-8-1987. (vide DD-6, PP. 190-191)
89. 26-3-1988: Anti-rain experiment was performed from 26-3-1988 to 31-4-1988 at Mannu Ghat, Jarul Chhara, North Tripura to keep the clouds and rain away from the place of experiment and within the perimeter of 150 to 200 km. It was arranged by the contractor, Priyavrata. (vide DD-6, P. 82).
90. 23-11-1989: Rain-experiment was done from 23-11-1989 to 24-11-1989 at Farmana in order to facilitate the seeding of wheat. (vide DD-6, P. 101).
91. 30-10-1989: Rain-experiment was performed from 30-10-1989 to 4-11-1989 with the guarantee of rain up to 10-11-1989 at Anwali, Riwara, Mazra and Farmana. The rain-fall was witnessed on 4-11-1989 in varying degrees (i.e. from low to heavy) at the places of experiment and closing areas. (vide DD-6, PP. 83-94).
92. 5-11-1989: Rain-experiment were carried out from 5-11-1989 to March 1990 in order to make the weather favourable to the seeding and growth of crops all through the targeted period and it was a success. (vide

DD-6, PP. 88-130).

93. 11-5-1990: Rain-experiment was carried out from 11-5-1990 to 30-5-1990. Heavy rains were reported on 26-5-1990, 27-5-1990 and 30-5-1990 from all over North India. (vide DD-6, PP. 137-141).
94. 1-6-1990: Rain-experiment was done from 1-6-1990 to 30-6-1990 at Farmana. It rained heavily up to 30-6-1990 and so it had to be stopped to help the farmers sow the millet (*Jawar* and *Bajra*) easily. (vide DD-6, PP. 140-146)
95. 1-7-1990: Anti-rain experiment was carried out from 1-7-1990 to 13-7-1990 in view of a heavy downpour of rain caused by the preceding rain-experiment. Success was achieved and there was no rain within the perimeter of 100 km. of the place of experiment. (vide DD-6, PP. 147-155).
96. From 1-8-1990 to 31-8-1990 it so experimented as to let the reasonable amount of rain-fall occur in the region and so floods were checked accordingly. (vide DD-6, P.153).
97. From 1-9-1990 to 31-10-1990 a good amount of rain was allowed with necessary checks over it whenever so required. Thus a better seeding of crops was facilitated. (vide DD-6, P. 153).
98. From Nov. to March 1991, a reasonable amount of rain was induced, in order to facilitate the better growth of crops in the area. (vide DD-6, P. 153).
99. From April 1991 to May 1991, no rain was allowed to take place in order to save the wheat crop from anticipated damage by the rain. (vide DD-6, P. 154).
100. 20-7-01991 to 3-8-1991: Rain-experiments were conducted at G.H.S. Bichpari, Distt. Sonepat, Jharli, Distt. Rohtak; G.H.S.S. Kosali, Distt. Rohtak; Bhakali, Distt. Rohtak; G.P.S. Majra, Distt. Sonepat, Haryana, consequent upon which Bichpari proper had low rain whereas heavy rain was experienced by the

places situated outside Bichpari. Jharli recorded moderate rain-fall on 24-7-1991. Kosali and Bhakali had to go without rain, though the sky was overcast with clouds. Heavy rain was also witnessed in Majra and its surrounding area up to 20 km. (vide DD-6, PP. 155-156).

101. 22-11-1992 to 21-5-1993 Anti-rain experiment was performed in Bangalore city.

102. 22-5-1993 to 28-5-93 Rain experiment was carried out in Bangalore city. It rained intermittently for three days between May, 24 and 27.

103. 13-6-1993 to 10-7-1993: 'Rain *Yajña*' conducted at Rohtak 1051, Sector No. 1. Heavy Rain occurred.

104. 12-7-1993 to 31-8-1993: Anti-rain *Yajña* to check the flood was carried out with the help of Rohtak residents.

105. 1.9.93 to 12.9.93: Rain *yajña* was conducted in Rohtak. Consequent upon which medium rain was experienced in the close-by area.

106. 17.10.93 to 31.10.93: Rain *yajña* was performed at Rohtak with the help of the residents of Rohtak. The result was heavy rain at proper Rohtak and surrounding area.

107. 1.4.94 to 31.5.94: Anti Rain *yajña* was carried out to allow the safe harvest of crops.

108. 2.5.95 to 31.5.95: Anti-rain *yajña* was carried out to save the harvesting of crops from being spoiled by rain.

109. 8.6.95 to 21.6.95: Rain *yajña* was carried out in village Muradpur Tekna with the help of Jai Narain Sangwan, Sarpanch of the village. Medium to heavy rain was reported by Newspapers.

110. 22.6.95 to 26.6.95: Anti Rain *yajña* was performed in Rohtak to facilitate the local farmers disruption-free harvesting of crops.

111. 31.7.95 to 4.8.95: Rain *yajña* was conducted at village

Siwana, Dist. Jhajjar with the help of village panchayat. Heavy rain occurred in the area. This experiment was reported in many local papers.

112. 5.8.95 to 27.8.95: Anti Rain *yajña* was conducted with the help of Sunder Shyam Malhotra and Distt. Administration and Red cross society. The rain was prevented from occurring around the periphery of 60 miles of Rohtak. As and when the experiment was stopped, the city was inundated with heavy rain. Thereafter up to 3.8.95, there occurred heavy rain.

113. 4.9.95 to 30.9.95: Anti Rain and medication *yajña* had to be performed with the help of Dharam Sukh Dahiya, Principal Evening college Rohtak, to save the city of Rohtak from the heavy downpour and prospective outburst of the epidemic. The rainfall could be prevented within the periphery of 60 km. of Rohtak.

114. 24.5.96 to 13.6.96: Rain *yajña* was performed at Rohtak and village Bohar to facilitate the seeding of crops. The medium rain was reported throughout Haryana, Panjab and Delhi.

115. 15.7.96 to 19.7.96: Rain *yajña* was carried out at village Siwana, Distt Jhajjar with the demand of villagers.

116. 29.7.97 to 1.8.97: Rain *yajña* was conducted at the request of the villagers of Siwana. Heavy rain was reported from the place to *Yajña* and the surrounding area of 60 km.

117. 22.8.97 to 26.8.97: Rain *yajña* was conducted at village Killoi with the help of villagers and Master Hari Singh Hooda. Heavy rain was reported in the area. The District administration acknowledged the success of the Rain *yajña.*

118. 9.9.97 to 18.9.97: Rain *yajña* was conducted at the request of the villagers of the village Maharana, Distt. Jhajjar which resulted in drizzling at the place of *Yajña* and medium rain in the surrounding area of 60

km.

119. 19.9.97 to 30.9.97: Rain *yajña* was performed at the village of Khanpur Kalan (Gohana), Distt. Sonepat with the help of Chattar Singh, the Sarpanch of the village and other villagers. Place of *Yajña* witnessed drizzling and the surrounding area of 100 km. had medium rain. This experiment was reported in '*The Dainika Tribune*' '*The Hindustan Times*' and a local paper '*Mudrika*'.
120. Between 1.2.95 to 1.8.99: Anti Rain *yajña* was performed continuously at the village of Muradpur Tekna during the medical camp to be organised on second Sundays of the months for the smooth conduct of the camp. These experiments were conducted at the request of Jungli Pahalwan. There occurred rain on four occasions which was warded off immediately by the experiment.
121. 3.6.98 to 6.6.98: Rain and Anti-rain *yajña* was tested by Swami Jiwananda, registrar, Gurukul Jhajjar at Jat Bhurthal, Distt. Rewari, Haryana. During this rain experiment. A particular area of Brick kiln was saved from being lashed out by the downpour of rain conducted through Rain experiment.
122. 8.6.98 to 20.6.98: Rain *yajña* was conducted at many places in Rohtak. Moderate rain was witnessed by the city from 13.6.98 to 19.6.98. This was reported by '*Dainika Bhaskara*' a local newspaper of the area.
123. 26.7.98 to 17.8.98: Rain *yajña* was conducted at village Siwana and Rohtak proper. Newspapers reported heavy rain from 9.8.98 to 22.8.98 in the surrounding area, whereas the place of *Yajña* had to content with moderate rain. This experiment was reported in '*Haribumi*', a local newspaper on 31.7.98 and 16.8.98.
124. 1.4.99 to 15.5.99: Anti Rain *yajña* was conducted to save the crops from the prospective rain.
125. 4.5.2000 to 15.5.2000: Rain *yajña* was performed at

Pathmeda Gaushala, Rajsthan. There occurred drizzling at the place of *Yajña.* The area from 15 to 300 km. witnessed heavy rainfall. This was published in '*Rajsthan Patrika*' on 14.5.2000; '*Aas Pas Patrika*' (Jodhpur) on 8.5.2000; and '*Rajmanch Patrika*, Jodhpur on 12.5.2000. Moderate to heavy rain was reported throughout India from 8.5.2000 to 15.5.2000.

126. 28.5.2000 to 2.6.2000: Rain *yajña* was performed during Arya Samaj annual function at Mount Abu, Rajsthan. There occurred drizzling at the place of *Yajña* and moderate to heavy rain was reported from the surrounding area.

127. 3.6.2000 to 15.6.2000: Rain *yajña* was performed at Red cross society, Rohtak and Govt. Book Depot Rohtak. Courtesy Jainarain Gahalawat and Bhale Ram Arya. This experiment was reported in '*Dainika Bhaskar*' on 12.6.2000 and '*Dainika Tribune*' on 15.6.2000. Heavy rain was reported throughout North India including the place of *yajña* from 3.6.2000 to 12.6.2000.

128. 16.6.2000 to 30.6.2000: Anti Rain *yajña* was conducted to facilitate the farmers rain-free seeding of crops.

129. 7.7.2000 to 15.7.2000: Rain *yajña* was conducted in Rohtak with the help of many organisations and eminent persons. It was reported in '*Dainika Bhaskar*' on 12.7.2000 and '*Hari Bhumi*' on 9.7.2000.

130. 3.8.2000 to 14.8.2000: Rain *yajña* was carried out in the village Siwana at the request of the villagers. Moderate to heavy rain was reported in that area.

131. 2.10.2000 to 16.10.2000: Rain *yajña* was performed at village Khudan Chapara, Distt. Jhajjar. Place of *Yajña* witnessed drizzle drops only, rest of the area experienced 8- to10 centimetre of rain.

132. 19.10.2000 to 2.12.2000: Rain *yajña* was conducted with the help of various organisations in Rohtak.

133. 3.1.2001 to 24.1.2001: Rain *yajña* was conducted at various places in Rohtak. Moderate rain occurred in Rohtak and surrounding area.

134. 20.4.2001 to 14.5.2001: In view of the damage to the crops due to the heavy downpour throughout the Northern region from 1.4.2001 to 19.4.2001, an anti-Rain operation was carried out from 20.4.2001 to 15.5.2001 to prevent unwanted rain and to save crops from being ruined due to the excessive rain. This was done with the help of the public, various institutions and companies at the following places: JNV Kiloi Jhajjar, Rohtak (20-21 April), IFFVS Sector-1, Rohtak (22-23 April; 25.4.2001 to 4.5.2001; 6 to 9 May and 11,13 & 14 May), Maruti Car Agency, Jagmohan Mittal, Rohtak (24 April), Sector-14 Kailash Verma, Asst. Director, Radio Station Rohtak (5 May), Village Jasrana Dist. Sonepat (10 May), Housing Board, Sector-1 (12 May). On 14.5.2001 at Gohana (Rohtak). The operation remained successful. There was no rain during this period. The operation was also reported by '*Hari Bhumi*', a daily Hindi paper on 19.5.2001.

135. 15.5.2001 to 15.6.2001: In view of the demands of the farmers of the area, operation for rainmaking was conducted. This operation was carried out with the help of various agencies at the following places: at IFFVS Rohtak (15.5.; 16.5, 28-30.5; 2.6-15.6.2001), Mahindra School Rohtak with the help of Principal Shashi Kala and Sandeep Chaudhary (17.5.2001), Govt. Book Depot (18.5.2001) with the help of Bhale Ram Arya and others, Jat High School Rohtak with the help of Principal Rambhaj (19.5.2001); Krishan Nandal and Ram Kumar Dalal (20.5.2001; Forest Dept. Rohtak with the help of Mahender Singh (21.5.2001); L.N. Dahiya, ex. Pro-Vice-Chancellor, MD University, Rohtak and MDH School; Mata Darwaja Rohtak (22.5.2001) with the help of the Principal Sumitra Arya, Jagdish Chand Nagpal, Model

Town, Principal Raj Bal & Raj Kumar Adv. Rohtak (24.5.2001); Devender Varma, Janta Colony, Rohtak (25.5.2001); Baljit Dahiya & Raj Kumar Secretary, Ram Gopal Colony, Rohtak (26.5.2001); Yajna Datta Arya & Charan Singh of Arya Colony; Rohtak (27.5.2001), Jai Narain Gahalawat, Red Cross Secretary, Rohtak (31.5.2001); Co-operative Management Centre Rohtak with the efforts of Principal Zile Singh (1.6.2001). As a result of this operation, there was heavy rain throughout U.P., Delhi, Haryana and Panjab. This operation was also reported in the Hindi daily '*Hari Bhumi*' on 19.5.2001.

136. 1.7.01 to 21.7.01: Anti-rain *Yajña* was conducted from 1.7.01 to 21.7.01 at IFFVC Rohtak to relieve the farmers from excessive rain.

137. 23.7.01 to 27.7.01: Rain *Yajña* was performed at Rohtak from 23.7.01 to 27.7.01 as a result heavy rain was recorded at Rohtak and surrounding area on 26.7.01 and 27.7.01.

138. 1.8.2001 to 31.8.2001: In view of the demands of the farmers of the area, operation for rainmaking was conducted in the month of August at the office of Distt. council Rohtak (from 1.8.2001 to 8.8.2001) due to courtsey to Dharma Pal Makrauli, Chairman, Distt. council. From 9.8.01 to 21.8.01 at Village Kiloi (Rohtak), resulting upon which, proper Kiloi and surrounding area witnessed moderate to heavy rainfall. At Ram Gopal Colony, Rohtak (22.8.01) courtesy Rajkumar Siwan; at IFFVS (22.8.01 to 23.8.01); at Timarpur Delhi (24.8.01 to 29.8.01); at IFFVS (30.8.01 to 31.8.01). As a result, heavy stormy rain was recorded on 25.8.01. This *Yajña* was reported in '*Hari Bhumi*' on 2.8.2001.

138. 13.9.01 to 30.9.01: *Yajña* for cloud formation and to check diseases was conducted at Rohtak, courtesy Mr. I.S. Dalal, Principal Jat College Rohtak, Mrs. Rita

Hooda, Principal Mahindra Model School Rohtak, Satbir Hooda Advocate, Suresh Arya Publisher, Cpt. Randhir Singh Badhwar, Krishan Nandal, Welder Rohtak and Karan Singh, the village Head, Bhainswal Kalan Sonepat.

139. 1.10.01 to 11.10.01: Cloud seeding operation was carried out with the help of '*Yajña*' performed at Rohtak.

140. Jan.-March 2002: *Yajñas* were performed at various places in Rohtak and Faridabad cities of Haryana and Delhi (14.2.02 to 21.2.02) for seeding clouds and preventing diseases on account of changing weather conditions in northern India.

141. 15.5.2002 to 31.5.2002: A cloud seeding and rainmaking *Yajña* was carried out in Rohtak, Haryana under the aegis of IFFVC. This *Yajña* was also reported in a Hindi daily Newspaper '*Hari Bhumi*' on 16.5.2002 and 29.5.2002. Following people from Rohtak involved themselves in carrying out this operation: Jagmohan Mittal of Maruti Car Agency, Dharam Vira Nandal of sector-14, Ramkumar Dalal, Surender Varma, Raj Kumar, Ishwar Singh Dahiya, Raj Singh Hooda JE, Mitter Sen Sindhu, Ram Prakash Arya, Kapoor Singh-Retd. Deen of Colleges MD University Rohtak, Ashoka Kadyan, Mahender Singh Hooda-Principal Jyoti Prakash H/S sector-1, Estate Officer HUDA, Yeshender Singh SDM, Suresh Arya of Saini Pura, Prem Singh, LN Dahiya-Ex. Pro Vice-Chancellor MD University Rohtak, Consequent upon which heavy rain took place in Haryana, Punjab, Delhi and Rajasthan.

142. 16.6.2002 to 30.6.2002: Second rain *Yajña* was conducted with the involvement of Shri Krishan (Principal of Chotu Ram college of Education) Mahender Singh Chhahar, Dayanand Lohvan (MD University), Ashok from Panjab National Bank, Jitender Varma from Janta Colo ny, Rathi SDO of

HUDA office, Surender Varma, Ram Chander Varma and Devender Varma of Janta Colony Rohtak. This *Yajña* was also reported in the Hindi daily '*Hari Bhumi*' on 18.6.2002.

143. 11.7.2002 to 13.9.2002: In the year 2002 northern parts of Bharat suffered from a severe drought. The entire rainy season went without rain. To tackle the situation of drought in northern India (Bharat), Indian Foundation for Vedic Science carried out a rainmaking operation from 11.7.2002 to 13.9.2002 in Rohtak and Jhajjar Districts of Haryana with the participation of various organisations and Individuals which are listed below: Arya Samaj Shivaji Colony (11.7.02 to 21.07.02 which was reported in *Hari Bhumi*, a local daily on 9.7.02); Naresh Raja Electronic, Jhajjar Road; Inder Lal, Adarsh Nagar; Principal (S.S. Arya) University Campus School; Subhash Bhalla, President of Hanuman Mandir Chinaut Colony; Satish studio (Delhi Road); Satyender (Delhi Road); Nilakantha Mandir (Patel Nagar); Dharambir Nandal (Sector-14); Arya Samaj, Model Town; (18.7.02 to 28.7.02 which was reported in local dailies named *Hari Bhumi* and *Dainika Bhaskar* on 20.07.02); Jaisingh Ahalawat, (Sector-1); Rajbir (Kheri village); Raj Pal Rathi (SDE, HUDA office); Sardar Lakhminder Singh, JE, HUDA Office; Sanatan Dharma Mandir (Ward no. 3 Camp); Ishwar Singh (Sundana village); Dharam Pal Makarauli, (Chairman, Office of Jila Parishad); Suresh Huda, Ram Singh Kadyan, Vijaya Singh Chohan and Satish Kumar Dhul from Rohtak (26.07.02 to 16.08.02); Ram Bhaj Huda (Principal, Jat High School Rohtak); Arya Samaj of Mokhara village (4.08.02 to 11.08.02); Mahender Singh, Pt. Ramkishan, Ramchandra Shyami, Vijayender, Dhuppal, Satbir, Naresh Arya, Kaptan, Subhash and Hoshiar Singh of Village Siwana, District Jhajjar (16.08.02 to 03.09.02); Consequent upon which heavy rain took place lashing out the maximum region of

north India. This is the first time that these parts of India witnessed such a heavy rainfall in the post-monsoon season.

144. 14.9.2002 to 30.9.2002: *Yajña* was conducted for medication and to check the flood situation likely to arise out of the rain operation conducted earlier from 11.09.02 to 13.09.02. Following persons/Institutions participated in this operation. Raj Kumar, Secretary Co-operative Society; Balbir Sing Dahiya, Advocate; Ram Bhaj Huda, Headmaster Jat High School; Raj Singh Nandal, President Jat Institutions in Rohtak; Rajender, Sarpanch of Siwana village; Mahender Singh Chhahar, Principal of Jat College of Education Rohtak. Mahendra and Sahil Dairy.

145. October-November and December 2002: *Yajña* was performed in Rohtak Distt of Haryana for cloud seeding, precipitation of rain, and medication (germ-free) atmosphere. Following persons led by Sh Mitra Sen Sandhu contributed to the success of *Yajña.* Names of other contributors are Rameshwar Dass (ASI Gohana-Sonepat), Dr. Pannu (Gohana, Sonepat), Rajkumar (Siwana Jhajjar), Naresh Bhagat (Delhi Road, Rohtak), Sh. Mitra Sen Sandhu, Principal Mohinder Singh Chahar (C.R. College of Education Rohtak), Dharmender Singh (Housing Board, Sector-1, Rohtak), Yaj Datta Arya (Arya Nagar, Rohtak), Satish Studio (near Bajrang Bhawan), SS Arya (Principal, M.D.U. Campus School Rohtak), Krishan Nandal (Welder - Rohtak), Surender Varma (HUDA Rohtak), Baljeet Dahiya (Typist, Distt. Courts Rohtak), Raj Bala (Principal, C. R. Public School Rohtak), Consequent upon this *Yajña*, the rain took place on 7.10.2002, 10.10.2002, 9.12.2002, 17.12.2002, 25.12.2002 to 27.12.2002 at various parts of North India.

146. 23.3.2003 to 29.3.2003: Rain-*yajña* was performed at Dehradun, Uttranchal. The city witnessed drizzling for two-three days.

147. 23.11.2003 to 25.11.2003: Rain-*yajña* was performed at Degana, Rajasthan. There occurred drizzling on 24.11.2003, although the sky remained overcast with clouds for all the days of *yajña*.

148. Between 1.4.2003 to 30.6.2003: *Yajña* for healing, cloud-formation and rain was carried out from time to time with the participation of various organisations and Individuals which are listed below: Anaj Mandi Rohtak Through Rakesh and Hansraj on 1.4.2003; Haryana Public school arranged by Principal J. S. Bura on 2.4.2003 (persistence of clouds continued); village Balyana Rohtak arranged by Dharam Veer Varma on 10.4.03; Old Anaj Mandi Rohtak on 11.4.03; Arya Samaj Shivaji Colony Rohtak on 20.4.03; Faridabad Sector 17 from 10.5.03 to 11.5.03; M. D. University Campus School arranged by Principal S.S. Arya on 17.5.03; Mohindra English medium Model School arranged by Principal Smt. Rita Hooda on 21.5.03; Jag Mohan Mittal at Maruti Car Agency Office on 24.5.03; Deptt. of Geohydrology arranged by N. K. Khattar on 26.5.03; Sector 4, Rohtak arranged by Shri Krishan Hooda and Smt. Rita Hooda on 27.5.03; on 7.6.03 it was arranged by Krishan Nandal and Yajña Datta Arya from Rohtak; on 8.6.03 by R.R. Banswal Commissioner Rohtak Division; on 9.6.03 by Shri Ram Chandra Varma and Devendra Varma from Janta Colony Rohtak; on 10.6.03 by Sh. Mehar Singh and Ravi, coaches Wrestling Rohtak; on 14.6.03 by Sh. Surinder Varma Rohtak; on 15.6.03 by Sh. L.N. Dahiya ex Pro-Vice-Chancellor Rohtak; on 29.6.03 by Arya Samaj Shivaji Colony Rohtak; on 30.6.03 by J.S. Bura, Principal, Haryana School Rohtak. Consequent upon these *yajñas* heavy clouds were formed and persisted throughout, the month resulting into drizzling on 3.4.03 and medium rain on 16.4.03; 19.4.03; 26.4.03; 27.4.03; 13-15.5.03; 19&22-25.5.03; 1.6.03; 18-20.6.03 and 21,23,24,26,27,& 29.6.03.

149. 8.1.2004 to 22.1.2004: A Rain-*yajña* was performed at Arya Samaj Indira Colony. Brahmachari Satyakam and Om Prakash helped organise this *yajña.* Consequent upon the *yajña,* heavy rain occurred in the city proper and the surrounding area.

151. 23.7.2004 to 4.8.2004: A Rain-*yajña* was performed involving many Institutions like Haryana Public School, All India Radio, Jat High School, Jyoti Prakash School, Arya Samaj Shivaji Colony, Yoga Ashram, MDU Campus School, Kishori mal College, and Philanthropists like Yagya-Dutt Arya, Pritam Singh, Shri Kishan Sharma, Baljeet Dahiya, Principal Krishna Chaudhary, Headmaster Rambhaj Makrauli, Principal K.S. Bura, Principal S.S. Arya, Prof. Karan Singh Rathee, Jagmohan Mittal, Devender Baharat, R.C. Varma, Bhim Singh Sangwan and Advocate Prashant. As a result of the *yajña* sky remained overcast with clouds and the areas of in the periphery of 100 km. of *Yajña* witnessed a moderate to heavy rainfall.

152. 9.9.2004 to 31.10.2004: Rain-*yajña* was performed in Rohtak at many places. Moderate to heavy Rain occurred throughout Haryana.

153. The state of Andhra Pradesh was suffering from scarcity of rain for the last 5-7 years. Due to the scarcity of rain, farmers in AP state were committing suicide. Govt of AP was also resorting to the technique of cloud seeding with the experts in this area. In the year 2004, the Govt. of AP spent around 20 crore Rupees on the cloud seeding on this project. As per news appeared in Deccan chronicle (11.02.05), the conservative estimate of the Govt. of AP was that success rate of cloud seeding experiments was 10 percent, resulting in 8 cm. of rainfall. It was also not surely be said that whether the rainfall was only due to cloud seeding. Now the State Govt. has set up a panel for cloud seeding to experiment with it for another

four years. The state Govt. is in a process of preparing a proposal of Rs. 100 crores in *collaboration* with neighbouring states and the Union Govt. reported *The Hindu* on Feb 6, 2004. In view of the same grievous situation, Kaktiya Cement Sugar Mills Ltd. Hyderabad invited IFFVS to perform rainmaking operation just to help the famine-stricken people with the rain made through the Vedic conventional way. At the invitation of Kaktiya Cement and Sugar Mills Ltd. Hyderabad, Ram Narain Arya, Chairman IFFVS reached Hyderabad on 13.12.04 to take stock of the climatic conditions there. He conducted some rainmaking operation at Peddamberpet, Hayathnagar Mandal, Ranga Reddy District adjacent to Hyderabad from 13th to 21st Dec 2004 with the Havan material available there. But it was found that the material available in Hyderabad was not so effective in inducing rain. So, he came back to Delhi on 22.12.04, collected material for Rs. 10680 and went back to Hyderabad on 25.1.05. He started his operation from 26.12.2004. From 28.12.04 clouds started forming and persisted through day and night. On 1.1.05 even the sun was not visible at the site of *Yajña*. On 3.1.05 slight rainfall took place in the close by areas. The weather remained cloudy till 10.1.05 the time when the havan material exhausted. On 14th the consignment of new Havan material purchased at the cost of Rs. 13000 from Delhi was received. Thus the second phase of the experiment started on 14.1.05. From 22.1.05 isolated rain took place at the place of *Yajña* as well as surrounding areas. *The Hindu* (23.1.05) reported 3 cm. of rain at Tuni and 1 cm. at Vishakhapattanam. The rain continued on the next day also. From 25.1.05 to 5.2.05 isolated to moderate rainfall occurred throughout Andhra Pradesh. The Hindu (25.1.05) reported that there was isolated rain in Telangana and Rayalseema, Kurnool and Nalgonda recorded a rainfall of 1 cm. each. The Hindu, (27.1.05) reported that there was isolated rain in coastal Andhra and Telangana. Hyderabad and Medak received rainfall 2 cm. each and Vishakhapattanam 1 cm. *The Hindu* (28.1.05) reported isolated rainfall in coastal Andhra owing to upper air circulation extending to Telangana from neighbouring states, Narsipattnam recorded a rainfall of 2 cm. and Hakimpet near Hyderabad one cm. for the day. According to The Hindu (1.2.05), heavy rain lashed the twin cities on 31.1.05 night leaving the roads in several areas under sheets of water. The downpour continued for 30 minutes and the drizzle continued till late in the night. The Met. Dept. said that 5.9 mm rain was recorded till 11.30 p.m. Adilabad recorded a rainfall of 12 cm., Medak recorded 3 cm and Nizamabad 1 cm. *The Hindu* (2.2.05) reported that there was rain at many places in Rayalaseema. Bhadrachalam recorded a rainfall of 5 cm. while Hakimpet airport and Medak, Kadiri and Medhira recorded 2 cm. each. *The Hindu* (3.2.05) reported that Venkatagiri recorded heavy rainfall of 7 cm. on 2.2.05 and on 5.2.05 the same paper reported isolated rainfall occurring over Coastal Andhra Pradesh and Rayalaseema. In view of the success of rain, the third consignment of havan material amounting Rs. 20000 was again purchased from Delhi. The *yajña* continued till 22.2.05, but no rain could occur. The main reason was that the state Govt. following an international seminar on cloud seeding convened on 28.01.05 at Jawahar Lal Nehru Technological University, started cloud seeding experiments. So the cloud seeders started seeding clouds formed by the rainmaking operation. During the third phase, there was no rain. Moreover, it was found by the rainmaker that the People in Brick kilns were using such type of material as was preventing rain from falling. So long as the state Govt. do not take appropriate steps to prevent people in brick kilns not to burn that anti-rain material, the state continues to face the famine-like situation. Here it may also be pointed out that the state can receive rain at a very low cost through Vedic operation of rainmaking as compared to the highly expensive technique of cloud seeding, which has only a 10% success rate.

154. 26.8.2005 to 17.9.2005: Due to scanty rainfall, the meteorologists and astrologers declared drought in Rajasthan. Keeping in view of the water crisis, the villagers of Siana, Teh. Kolayat, District Bikaner headed by Sh. Ganga Ram and Ram Narain Syami of Bikaner invited Ram Narain Arya, Chairman of IFFVS for carrying out Rainmaking operation from 26.8.2005 to 17.9.2005 near Nainiya village, Teh. Kolayat, Bikaner and at village Sobhasar, Teh. Kolayat. The operation was started on 26.8.2005 in the village Siana. From 29.8.2005, the venue was shifted from Siana to the agricultural land of Sh. Ganga Ram adjoining Nainiya village, Teh. Kolayat, Bikaner. Consequent upon the operation, from 30.8.2005 onward the sky began to overcast with clouds and from 5.9.2005 onward rain started pouring in various parts of Rajasthan. On 6.9.2005 there occurred drizzling at the place of operation also. From 9.9.2005 to 14.9.2005 there was heavy rain at the place of operation (on 9th, 13th and 14th it continued for almost 24 hours). The news of rain was reported from various parts of Rajasthan and adjoining states. In Jhajhu, the rainfed Jhajhu river was inundated. The cities like Gangasahar, Bikaner, Ajmer, Udaipur, Jaipur witnessed flooded water. There was heavy rain not only in Rajasthan but in the adjoining states and throughout India. This news was also reported by the local newspapers "*Dainik Bhaskar*' dated 30.8.2005, and 14.9.2005. It was a successful operation. The farmers were very happy.

155. 11.10.2005 to 31.10.2005: In view of the heavy

success of earlier rain-operation, people from other parts of Rajasthan wanted to avail the expertise of IFFVS in the field of rainmaking and prevention of rain. The second rain-making operation was carried out in Gajner, Bikaner at the request of the villagers of Gajner headed by Sh. Naula Ram. The operation started on 11.10.2005. As a result of this operation, the sky became overcast with clouds the next day itself. The people around the area were happy to see the signs of rain. Local newspaper '*Rajasthan Patrika*' confirmed the impact of Rainmaking operation on 16.10.2005. Although the sky remained overcast with clouds and many other parts of India witnessed heavy rainfall, but almost all the parts in Rajasthan remained dry.

156. 30.12.2005 to 2.1.2006: Rain-*yajña* was performed at many places in Rohtak with the assistance of Manish, Principal M.S. Chahar. There occurred drizzling on 2.1.2006 at the place of *yajña* and surrounding area witnessed moderate rain.

157. 28.2.2006 to 20.3.2006: Rain-*yajña* was performed successfully at various places in Rohtak involving Mehar Singh's Akhara; Jat High School; Haryana Public School and eminent social workers Devender Bharat, Yagyadatta Arya, Dharmendra, Rudh Sing, S.S Sangwan (Director Medical college), Khattar, Smt. Krishna Chaudhary (Principal Jat Girl's college), Jora Sarpanch, Maman Singh, Manjeet Rathi, Hawa Singh and Surinder etc. There occurred moderate to heavy rain in Rohtak and surrounding areas. This experiment was reported by a local daily newspaper '*Sāndhya Danika*' on 6.3.2006, 11.3.2006 and 16.3.2006.

Newspaper Reports

Following news regarding the experiments of rain and Anti-rain were reported by News-Papers from time to time.

1. Oct. 1972: The *Rajadharma,* a weekly paper, published under the head '*Yajña varṣā*' an ad. reporting guaranteed rain with the help of *Yajña* performed by Shri Ram Narain.
2. 31-12-1972: The *Rajadharma* again reported the news of inducing rain with the help of *Yajña* in a form of challenge that if there was no rain within the perimeter of 60 miles of the place of *Yajña,* expenses incurred upon *yajña* will be returned summarily.
3. Nov. 1976: Arya Samaj in its centenary commemoration volume recognised Ram Narain as a rainmaker and hence reported him as a specialist in rain-formation.
4. 10-7-1983: 'The Arya Jagat' from New Delhi reported the rain-induced by the Rain-experiment of *Gurukul Indraprastha* conducted from 5-6-1983 to 12-6-1983.
5. 29-2-1984: 'The *Rajadharma*' reported Ram Narain's claim of inducing rain with a guarantee at the cost of Rs. 5000 and Rs. 200 per day for preventing the same. In case rain or prevention of rain didn't take place, money incurred on the experiment would by returned summarily.
6. March 1984: 'The *Rajadharma*' reported the experiments of Anti-rain carried out from 21-3-1984 to 31-5-1984 and also that of rain performed from 1-6-1984 to 7-6-1984.
7. March 1984: 'The Vir Arjun' reported Ram Narain's experiments of Anti-rain performed from 21-3-1984 to 31-5-1984 and likewise the experiments of inducing rain performed from 1-6-1984 to 7-6-1984.
8. 26-4-1984: 'The *Janasatta,*' a Hindi daily from Delhi, reported rainmaker's experiment of rain as '*Yaga se*

barsāt', i.e. rain with the help of *Yaga.*

9. 28-4-1984: 'The *Sarvahitkari'* from Rohtak, Haryana also reported this experiment of inducing rain under the head *Yaga se barsāt,* i.e. Rain with the help of *Yaga'.*
10. 21-6-1984: 'The *Sarvahitkari'* from Rohtak, Haryana reported the claim of guaranteed rain under the column *Ārya Samāj kī gatividhiyān* 'The activities of Arya Samaj'.
11. 30-6-1984: 'The *Sarvahitkaari'* from Rohtak, Haryana, reported the reiteration of rainmaker's claim of inducing rain at the proclamation of famine in India by the American Scientists.
12. 12-7-1984: 'The *Punjab Kesari'* from Delhi reported the claim of inducing rain under the head '*Yajña se varṣā'.* 'The rain formation with the help of *Yajña'.*
13. 16-8-1984: 'The *Rajadharma'* reported Rain-*yajña* performed at Govt. High School Rabharan from 8-8-1984 to 15-8-1984 under the column *Ārya Jagat kī Halchal.* 'The activities of *Ārya Jagat'.*
14. 10-10-1985: 'The *Foolchapa,'* a Gujarati paper reported the *Vṛṣṭi-Yajña* performed in Kasturbadham, Tankara, Gujrat.
15. 26-4-1986: *'The Indian Express'* from New Delhi reported the claim of inducing rain with the help of Vedic *Yajña* under the head 'Rs. 5000 for a rainy day'.
16. 16-6-1986: 'The *Foolchapa,'* a Gujarati paper reported the rain experiment conducted in Arya Samaj Mandir, Hathikhana, Rajkot.
17. 18-6-1986: 'The *Janasatta,'* another Gujarati paper reported rain formed with the help of *Vṛṣṭi-Yajña* conducted in Rajkot.
18. 19-6-1986: 'The *Loksatta,'* yet another Gujarati paper reported the Rain-*yajña* performed in Tankara, Gujrat.
19. 21-6-1986: 'The *Jayahind,'* one more Gujarati paper reported the rain experiment conducted in *Upadeshaka Mahāvidyālaya* Tankara, Gujrat.

20. 21-7-1986: 'The *Sarvahitkari'* from Rohtak, Haryana reported the success of rain experiments performed in Rajasthan and Gujrat with the help of special material prepared on the lines of the Vedas for the purpose.
21. 19-8-1986: '*Kachcha Mitra,'* a Gujarati daily, reported the rain experiment conducted at Nirena (Kachchabuj) Gujrat.
22. 1-3-1993: '*Dhamakranti'* from Bangalore reported the Rainmaker's views on the Vedic science of *Yaga* and *Yajña.*
23. 2.6.1993: '*The Hindustan Times'* from Bangalore reported the rain experiment for three days between May 24 and 27.
24. 6-10-1993: '*Rāstrīya Saharā'* from Delhi reported Rainmaker's claim to induce and ward off rain with the help of *Yoga* and *Yajña.*
25. 13-10-1993: '*Danika Hind Times'* from Rohtak reported the success of Rainmaker in his experiments of Rain and Anti-rain. The paper also published the weather forecast made by the Rainmaker for the next 5 days.
26. 22.6.95: '*Haryana Nirmana'* from Rohtak reported the success of Anti-rain-*yajña* performed at Muradpur Tekna from 22.6.95 to 26.6.95.
27. 24.6.95: '*Danika Hind Time'* from Rohtak reported the success of Anti-rain-*yajña* performed at Muradpur Tekna from 22.6.95 to 26.6.95.
28. 2.8.95 & 6.8.95: '*Danika Hind Time*' from Rohtak reported the rain-*yajña* conducted at Siwana from 31.7.95 to 4.8.95.
29. 14.8.95: '*Sarvahitkari*' from Rohtak reported the rain-*yajña* conducted at Siwana from 31.7.95 to 4.8.95.
30. 25.5.96: '*Danika Hind Time'* from Rohtak reported the success of rain-*yajña* performed at Bohar village from 24.5.96 to 7.6.96.
31. 11.1.1997: '*Bharat Janani'* from Rohtak reported the power of *yajña* to control the nature.

32. 22.9.97: '*Gohana Mudrika*' from Gohana reported the rain-*yajña* performed from 22.9.1997 to 27.9.1997 at Khanpur Kalan.
33. 23.9.97: '*Dainik Tribune*' from Rohtak reported the rain-*yajña* performed from 10.9.1997 to 18.9.1997 at village Maharana, Distt. Jhajjar.
34. 31.7.98 & 16.8.98: '*Hari Bhumi*' from Rohtak reported the success of the rain-*yajña* performed from 27.7.1998 to 17.8.1998 at Siwana village, Distt. Jhajjar.
35. 8.5.2000: '*Aas paas Patrika*' from Sanchor Rajasthan reported the rain-*yajña* performed from 4.5.2000 to 15.6.2000.
36. 12.5.2000:'*Rajya Manch Patrika*' from Jodhpur, Rajasthan reported the success of rain-*yajña* performed from 4.5.2000 to 15.6.2000.
37. 14.5.2000: '*Rajasthan Patrika*' from Rajasthan reported the rain-*yajña* performed from 4.5.2000 to 15.6.2000.
38. 12.6.2000: '*Dainik Bhaskar*' from Rohtak reported the heavy rain occurred due to rain-*yajña* performed from 3.6.2000 to 15.6.2000.
39. 15.6.2000: '*Dainik Tribune*' from Rohtak reported the heavy rain occurred due to rain-*yajña* performed from 3.6.2000 to 15.6.2000.
40. 22.6.2000: '*Dainik Bhaskar*' from Rohtak reported the anti-rain *yajña* performed from 16.6.2000 to 30.6.2000.
41. 9.7.2000: '*Hari Bhumi*' from Rohtak reported the rain-*yajña* performed from 7.7.2000 to 15.7.2000.
42. 12.7.2000: '*Dainik Bhaskar*' from Rohtak reported the rain-*yajña* performed from 7.7.2000 to 15.7.2000.
43. 10.8.2000: '*Dainik Tribune*' from Rohtak reported the success of rain-*yajña* performed at Siwana, Distt. Jhajjar from 3.8.2000 to 14.8.2000.
44. 12.10.2000: '*Danika Bhaskara*' from Rohtak reported the moderate rain due to rain-*yajña* performed from 2.10.2000 to 16.10.2000 at Khudan Chapar, Jhajjar.

45. 19.5.2001: '*Hari Bhumi*' from Rohtak reported the Anti-rain *yajña* performed from 20.4.2001 to 15.5.2001 and rain-*yajña* performed from 16.5.2001 to 27.5.2001.
46. 2.8.2001: '*Hari Bhumi*' from Rohtak reported the rain-*yajña* performed from 1.8.2001 to 8.8.2001 at Jila Parishad Rohtak.
47. 16.5.2002 & 29.5.2000: '*Hari Bhumi*' from Rohtak reported the rain-*yajña* performed from 15.5.2002 to 31.5.2002.
48. 18.6.2002: '*Hari Bhumi*' from Rohtak reported the rain-*yajña* performed from 16.6.2002 to 30.6.2002.
49. 9.7.2002 & 20.7.2002: '*Hari Bhumi*' from Rohtak reported the rain-*yajña* performed from 11.7.2002 to 13.9.2002.
50. 21.12.2002: '*Hari Bhumi*' from Rohtak reported the rain-*yajña* performed from 9.12.2002 to 30.12.2002.
51. 2.1.2003: '*Danika Jagarana*' from Rohtak reported the rain-*yajña* performed at Haryana Public School.
52. 27.2.2003: '*Rohtak Jagaraṇa*' from Rohtak reported the rain-*yajña* performed at M.D.U Rohtak.
53. 7.6.2003: '*Hari Bhumi*' from Rohtak reported the rain-*yajña* performed from 17.6.2003 to 17.7.2003.
54. 24 & 25.11.2003: '*Rajasthan Patrika*' from Rajasthan reported the rain-*yajña* performed from 23.11.2003 to 25.11.2003 at Digana Distt. Nagour.
55. 9.1.2004 & 13.1.2004: '*Hari Bhumi*' from Rohtak reported rain- *yajña* performed at Indira Colony Arya Samaj from 8.1.2004 to 22.1.2004.
56. 15.1.2004: '*Rohtak Bhaskar*' from Rohtak also reported the same rain-*yajña* performed from 8.1.2004 to 22.1.2004.
57. 1.8.2004: '*Hari Bhumi*' from Rohtak reported the rain-*yajña* performed from 23.7.2004 to 4.8.2004.
58. 4.8.2004: '*Rohtak Bhumi*' from Rohtak reported the rain-*yajña* performed from 23.7.2004 to 4.8.2004.

59. 1.10.2004: '*Hari Bhumi*' from Rohtak reported the rain-*yajña* performed from 9.9.2004 to 31.10.2004.
60. 30.8.2005 & 14.9.2005: '*Danika Bhaskara*' from Rajasthan reported the rain-*yajña* performed from 26.8.2005 to 17.9.2005 at village Siyana and Sobhasar, Distt. Bikaner, Rajasthan.
61. 11.10.2005 & 12.10.2005: '*Rajasthan Patrika*' from Rajasthan reported the rain-*yajña* performed from 11.10.2005 to 31.10.2005 at Gajner, Rajasthan.
62. 16.10.2005: '*Rajasthan Patrika*' from Rajasthan reported the effect of rain-*yajña* performed from 11.10.2005 to 31.10.2005 at Gajner, Rajasthan.
63. 21.10.2005: '*Rajasthan Patrika*' from Rajasthan reported the talk of the town regarding the rain-*yajña* performed from 11.10.2005 to 31.10.2005 at Gajner, Rajasthan.

Notes and References

1. Mason, B.J.: *Clouds, Rain and Rain-making,* 2nd Ed. Cambridge University press, 1975, P.124.
2. *VS.* 18.9.
3. *Ibid.* 22.22.
4. *Ibid.* 2.16.
5. *AV.* 5.24.5.
6. *VS.* 33.3.
7. *AV.* 3.109.
8. *Kath.* S. 30.10; *Kāth.* S. 46.8; *TS.* 3.2.1.4; *Ś. Br.* 1.7.1.5.
9. *Ś. Br.* 5.35.17.
10. *Bṛhadāraṇyakopaniṣad,* 6.2.10.
11. *Gītā,* 3.14.
12. *Ibid.* 3.15.
13. *Bṛhadvimānaśāstra,* 13.19.

14. *Ibid.* 22.48.

15. *Ibid.* 22.49.

16. *Ibid.* 22.50.

17. *Ibid.* 15.91.

18. *Ṛgvedādibhāṣyabhūmikā,* ed. Yadhiṣṭhir Mimāṁsaka, Bahalgarh, Sonepat, 1984.

19. All those who attempted rainmaking sought the help and guidance of Ram Narain Arya at times as regards the methodology and technique of inducing rain. Since their contribution to this field was not the original one, their names have not been included.

15

Method of Inducing Rain With the Help of *Yajña* or Rainmaking with the Help of *Yajña*

As it has already been stated that rainmaking with the help of *Yajña* is not an idea of recent origin, its genesis goes as far back as the composition of the Vedic *Saṁhitās.* The Vedic seer himself proclaims this scientific fact in one of his holy litanies as *vṛṣtiśca me yajñena kalpatām.*[1] 'Get me rain with the help of *Yajña.*' *Yajña* was so strong and effective a tool of the Vedic seer that it could be used to induce rain as and when desired. This is why, the *Ṛṣi* proclaims

'nikāme nikāme naḥ parjanyo varṣatu.[2]

'Let the cloud induce rain as and when we desire.'

According to the Veda, *Mitra* and *Varuṇa* are rain forming agents, e.g. *mitrāvaruṇau tvā vṛṣtyāvatām.*[3] 'May *Mitra* and *Varuṇa* bring rain for you.' *Mitrāvaruṇau vṛṣṭ yādhipati tau māvatām.*[4] 'May the rainmaking agents *Mitra* and *Varuṇa* protect you.' In fact, rainmaking with the help of *Yajña* was a systematic process of coordinating the rainmaking agents, the *Mitra* and *Varuṇa.* For instance, the *VS.* says :

yajñā no mitrāvaruṇā yajñā devaṁ ṛtaṁ bṛhat.[5]

'I.e. *Mitra* and *Varuṇa,* the main agents of rain should be coordinated with the help of *Yajña* for the precipitation of rain.

Mitra is a *soma* element or *āpastattava,* or say positive charge in modern scientific terms, *Varuṇa* is an *āgneya* element or *jyotiṣtattva,* i.e. a negative charge. Union of both the elements stimulates rain. *apāṁ ca jyotiṣaśca miṣrī bhāvakarmaṇo varṣakarma jāyate.*[6]. The deficiency of any of these elements in the atmosphere may be covered up with the help of their respective oblations (*āhutis)* offered in the terrestrial fire or *yajñīya* fire. Actually, *Yajña* acts as the remote controller and coordinator for the various deities, i.e. the natural elements existing in the mid-sphere or celestial sphere. Rainmaking with the help of *Yajña* is a function of *maitrāvāruṇī āhutis* offered to the fire in a particular ratio subject to the position of the two elements in the atmosphere. So, a rainmaker has a tough task before him. He will have to be very vigilant and constantly keep himself in touch with the weather conditions prevailing in the atmosphere. For rainmaking, he doesn't have to be, dependent, as the other meteorologists have to be, on the pre-existence of the clouds. In fact, he has to start from zero and modify weather, as the need be, to any such an extent as to blow moisture-bearing winds, to generate evaporation, to seed precipitation-producing clouds or to create other similar conditions leading to rainfall or simply to induce rain from the already existing clouds. This fact is also very well supported by the Brāhmaṇa writers' following observations regarding the precipitation of rain with the help of *Yajña.*

agnervaidhūmojāyate. Dhūmād abhram. Abhrāt vṛṣṭiḥ.[7]

That is to say, when *maitra āhutis* or *somīya āhutis* are offered to the *yajñīya* fire, smoke (the *vikāra* of waters existing in the mid-sphere) is produced, which, due to the law of specific elementary gravity, goes higher up in the atmosphere (*divaṁ te dhūmo gacchatu*[8]) through the fire column established with the help of celestial fire, the sun, in conjunction with the air. (*marutaṁ pṛṣatī gaccha, vaśā pṛsṇirbhūtvā divaṁ gaccha.*[9]) There it creates *abhra* (clouds) and clouds precipitate rain (*tato no vṛṣṭim āvaha*).[10]

Thus the rainmaker, with the help of *Yajña,* is capable of seeding clouds and inducing them to rain as and when required. But in the process, before starting a *Yajña* at any place or time, it is a priory of a rainmaker to make a proper survey of weather or study weather conditions properly. The weather is to be studied in terms of air-flow, i.e. whether the air draft is in a favourable direction or not; in terms of humidity or water vapours i.e. whether the air is humid enough to reach saturation point or dry requiring further evaporation of water from the ground, and in terms of cloudiness, i.e. whether the sky is seeded with sufficient clouds or there is no cloud at all. Thus a serious rainmaker should know the weather conditions properly and exactly before carrying out a rainmaking experiment. Different weather conditions would oblige rainmaker to modify weather differently for rainmaking and this would require different time spans and varied expenses to induce rain with the help of *Yajña.* This is why, Ram Narain Arya, a perfect rainmaker of modern times, in his challenges of guaranteed rain, the prohibition of falling rain, and other modifications pertaining to weather conditions requires different timings and costs of rain experiments subject to varied conditions of weather. To highlight this factual view, I am quoting here one of his pamphlets delivered on March 14, 1983, on the occasion of *Kṛṣi melā* organised by Pusa Institute of Delhi bearing the detailed programmes which could be demonstrated with the help of *Yajña.* The pamphlet reads thus:

nikām nikāme naḥ parjanyo varṣatu phalvatyo na oṣadhayaḥ pacyantām. yagakṣemo naḥ kalpatām.

Subject: Rain, Anti-rain, change of weather, pollution control, prevention of deluges, droughts, storms and disease through *Yajña.*

(A Research into Vedic Texts)

by

M. Ram Narain Arya,

M.A. (Sanskrit-Veda, Pol. Science)

V.P.O. Farmana, Tehsil & Distt. Sonepat, (Haryana)

The following programmes can be demonstrated in order to confirm the validity of the principles of the science of *Yajña.*

(Rain and Anit-rain)

1. The occurrence of rainfall at a definite place within 48 hours will be possible if the east wind is blowing and the sky is cloudy (day & night). The total cost would work out Rs.1000.
2. If the west wind is blowing and the sky is clear, it will take 7 days to make it to rain at a definite place (*yajñīya* place) costing at least Rs. 5000.
3. If the rain is falling, it can be stopped within 15 minutes of time of *yajña* at the cost of Rs. 200 per day.

(Change of Weather)

1. The required degree of temperature, humidity and rate of evaporation can be maintained in the atmosphere with the help of *Yajña.* Its cost would be the subject to the degree of temperature, humidity and evaporation to be maintained in the atmosphere.
2. East and west winds can be interchanged accordingly.
3. Timely or untimely arrival or departure of monsoons can

be regulated with the help of *Yajña.* In this way, deserts can be converted into fertile lands.

(Control of Air-pollution and Diseases)

1. *Yajña* can help control air pollution caused by industries' smoke, waste and warfares.
2. Several kinds of maladies pertaining to men, animals, plants and trees can be remedied with the help of *Yajña.*

All of this is due to my intensive studies, research and experimentation into the Vedic texts for the last 20 years. Now, methinks, this research requires proper care and help by the Govt. through any concerned institute, department or establishment, so that it can be applied for the well-being of humankind.

Dated 14 March 1983

Thus from the foregoing, it is crystal clear that the induction of rain or rainmaking depends exclusively on the current weather conditions. The weather may be dry cum hot; dry cum cold; humid cum hot; humid cum cold and accordingly air may be tropical (warm), polar (cold), maritime (humid) or continental (dry). With the help of *Yajña,* one can modify weather to the extent that it shall favour rainmaking. Following modifications of weather may be effected with the help of *Yajña* in order to induce rain.

Evaporation and Cloud-Formation

If the weather is dry and hot and the west-wind (continental air) is under sway, a rainmaker is required to do *vāyu-viloḍana* or *ākāśa-manthana,* i.e. an atmospheric stirring. For this, the air is heated up and made to flow around turbulently so that high evaporation of water from the ground or sea may easily be generated. See for example the following hemistich of the *RV.*

saṁ no iṣiro vātu vātaḥ.[12]

'Let the rain-precipitating winds blow from all directions.'

This is well-supported by the following verse of the *AV.*

prajāpati salilādā samudrādāpa iryanudadhimardayāti. udīryat marutaḥ samudratastaveṣo arko nabha ut pātyātha.[13]

> 'The sun assisted by air evaporates the terrestrial waters into celestial ones and makes them fall on the earth in the form of rain.'

The similar observation is made by the *Bṛhaddevatā.*

rasāna raśmibhirādāya vāyunāyaṁ gataḥ saha varṣatyeṣacayalloke tenendra iti smṛtaḥ.[14]

> 'The sun in conjunction with air conducts evaporation of fluid substances from the earth.'

For this type of heating and airflow *Vāruṇī āhutis* are offered to the *Yajña.* It augments the power of *Varuṇa* deity or the *āgneya* element in the atmosphere. Thus strengthened by *āgneya āhutis,* the *Varuṇ* element heats up the atmosphere and makes the air flow strongly. Two-three days of atmospheric stirring enables air to receive sufficient moisture required for saturation. Afterwards, condensation starts and clouds begin to form. But this is not sufficient to induce a good amount of rain. This is only the first phase of the process.

Persistence of Sea-bearing Winds

For sufficient rain, it is necessary for the moisture/rain/sea bearing winds and clouds to persist. In the context of Indian sub-continent, as it is well known, only the east-winds/monsoons bring moisture and rain. Hence, with the help of *Yajña,* the *somīya* substance is sprayed in the atmosphere to augment the *somīya* content of the air, which helps deflect the sea bearing winds/east-winds/monsoons.

Persistence of Clouds and Precipitation of Rain

For a cloud to persist, *Yajña-kuṇḍa* is maintained continuously day and night in order to keep the air ascending/slightly moving upward. Actually, when the *Yajña* is performed, the surrounding air is heated up and gets an upward thrust. This fact is also corroborated by the Vedic seer as. *svāhākṛte ūrdhavaṁ nabhasaṁ mārutaṁ gacchataṁ.*[15] 'When the *Yajña* is started, the air is accelerated upward.' This upward moving air, bearing the smoke of *somiya āhutis* offered to the *Yajñiya* fire, helps the clouds induce rain. This function of air is confirmed by the following authorities of *Vāyupurāṇa.*

mehgā vāyunighātena visṛjanti jalaṁ bhūvi.[16]

abhrasthāḥ prapatantyāpo vāyunā samudīritāḥ.[17]

Maitrāiṇī Saṁhitā also bears out the same fact, as *Maruto' mutaścyāvayanti, ete vai vṛṣṭyāḥ pradātāraḥ.*[18]

This is also well-supported by the following authorities of the *A V.*

marudbhih prachyutā meghā varṣantu pṛthivīmanu.[19]

marudbhiḥ pracyutā meghā saṁyantu pṛthivimanu.[20]

samabhrāṇi vātajūtāni yantu.[21]

upapruto marustāñ iryata vṛṣṭiryā viśvānivātaspṛṇāti.[22]

This process will take three to fifteen days to induce rain.

Secondly, there is likely that the favourable wind, i.e. east wind is already blowing and the sky is also cloudy. Under such conditions, a rainmaker is required to introduce such *somīya* contents in the atmosphere as could act as super-seeding agents. Clouds thus superseeded would be able to yield rain within 24 or more hours. This fact has been disclosed by the *Bṛhadāraṇyakopaniṣad* thus:

parjanyo vā'gnirgotama tasya saṁvatsara eva samidabhrāṇi dhūmo vidyudarcciraśanir aṇgāra hrādunayo visphuliṅgāstasminnetasminnaganu devāḥ 'somaṁ rājānaṁ juhvati tasyā āhutyai vṛṣṭiḥ saṁbhavati.

> 'That is to say, O' Gotam! *Parjanya* (the cloud) is the second sacrificial altar. It has a year as its *samidhā* (sacrificial fuel), *abhra* (supercooled cloud) as its smoke, lightning as its flame, *aśani* as its embre and the thunder

sound as its spark. This cloud can be induced to rain by way of augmenting the *somīya* element in the atmosphere with the help of *somīya āhutis.'*

Nature of Āhutis

The aforementioned discussion regarding the method of inducing rain with the help of *Yajña* remains incomplete so long as the nature of *somīya* and *vāruṇī āhutis* is not defined. Hence, a detailed break-up of *somīya* and *vāruṇī āhutis* in respect of various climatic conditions is rendered hereunder. To start with, it may be pointed out that for the sake of rainmaking or anti-rain experiments, climatic conditions should be studied into six types in the light of the variables of coldness and hotness in respect of humidity and aridity.

In this way the six climatic types to be worked out will be as:

1. maritime polar (humid cum cold),
2. maritime tropical (humid cum hot),
3. maritime temperate (humid cum equable).
4. continent polar (arid cum hot),
5. continental tropical (arid cum hot) and
6. continental temperate (arid cum equable).

Similarly, *āhuti dravya* (oblatory material) is composed of *auṣadhis* (vegetation) grown in various climatic conditions. From the point of growth of vegetation to be used as *āhutis* in humid or arid regions, the vegetations are classified as *somīya* and *āgneya* respectively and so are the *āhutis.*

Somīya Āhutis

As it is clear from the foregoing discussion, *somīya āhutis* are composed of *somīya* flora or vegetation growing in humid regions. They may be classified into three categories in respect to their growth in high, low or equable temperatures. For instance:

(1) *Somīya* or maritime flora growing in polar climatic conditions. The vegetation growing in low-temperature high humidity and with heavy rainfall may be classified as such. Evergreen oaks, chestnuts and chirpine trees are this type of vegetation which is located in upper Himalayan regions especially in eastern and southern zones.

(2) *Somīya* or maritime flora growing in tropical conditions. The vegetation growing in high temperature, high humidity with heavy rainfall in same climatic conditions is though hardly traceable in Indian peninsula, we have regions with slight variations i.e. with high temperature, comparatively low rainfall between 100 and 200 centimeters per annum, such as Foothills of Himalayas, Western ghat in south and Shivalika in the north. The vegetation growing in the above-named regions is known as tropical deciduous forests. Teak, sal, sandalwood, *shisham,* rosewood ebony, *mahua,* bamboos are this type of vegetation. Rice, coconut, sugarcane, tea and spices like black pepper, chillies, ginger and cardamom also grow under similar climatic conditions.

(3) *Somīya* or maritime flora growing in temperate conditions: The vegetation growing in even or moderate temperature, high humidity and heavy rainfalls under this category. This type of climatic conditions is found in Western Ghat, hills of Assam, West Bengal and Orissa. The evergreen and semi-evergreen forests growing in the above-mentioned regions belong to this category. Tea also belongs to this class of vegetation.

Vāruṇī Āhutis

1. Vāruṇī or continental flora growing in polar conditions. The vegetation growing in low temperature, aridity and

scanty rainfall belong to this category of flora. This type of climatic conditions may be traced to the northernmost area of Ladakh in Kashmir and Kinnaur and Lahul Spiti in Himachal Pradesh. The pines, fire and alpine variety of vegetation and other shrubs and grasses growing at the altitude of 2000 or above belong to this category. Herbs like *somalatā* are also of this type.

2. Vāruṇī or continental flora growing in tropical conditions. The vegetations growing under the conditions of high temperature, aridity and scanty rainfall may be categorized as such. Desert of Thar in India is the hot continent, so the shrubs, scrubs and thorny bushes growing in the desert of Thar are this types of vegetation. Moreover, horse gram and the millets or coarse grains like *Ragi, Jawar* and *Bazra* also belong to this category of vegetation.
3. Vāruṇī or continental flora growing in temperate conditions. The vegetation growing in moderate or even temperature, aridity and scanty rainfall falls under this category. Mustard and wheat are some examples of this type.

Here it may be clarified that vegetations always abound with such met. Qualities were prevalent in the weather during their growth. For instance, if particular vegetation grows in rainy conditions, it will also abound with rain promoting qualities. If a particular vegetation grows in the conditions of aridity, it will augment aridity. Similarly, the vegetations growing in polar or tropical conditions will augment polar and tropical conditions in the weather. The *Vajasaneyī* seer has clearly pointed out this fact in one of his verses as:

'varṣavṛddhamasi prati tvā varṣavṛddaṁ vattu.'[23]

'Since you have grown in rains, one should know you as rain promotor.'

While commenting on this verse, *Ś. Br.* exemplifies the rain-promoting agents like rice, reed and bamboo. For example, it had it as:

'atha śūrpam ādattle varṣavṛdhaṁ hyetadyadi naḍānāṁ yadi vetūnām yadīṣikāṇāṁ varṣamuhyevaitā vardhayati. '[24]

'He then, in the application of first part takes hold of a winnowing basket made of dried straw, reeds or bamboo. Since, straw, reeds or bamboo have grown in rain, promote rain'

Further in connection with the second part of the verse i.e.

prati tvā varṣavṛddhaṁ vettu, he maintains as 'atha havirnirvapati, prati tvā varṣavṛdhaṁ vettviti. varṣavṛddhā uhyevaiti. yadi vrīhiyo yadi yavā varṣ amuhya ivaitān vardhayati tatsaṁjñām eva etat śūrpāya ca vadati nedanyo'nyaṁ hinsāta iti'.[25]

'In the application of the second part, the oblation is prepared to be offered to the *Yajñiya* fire in the winnowing basket. The oblation is made of the rain promoting vegetation. Rice and barley grow in rain and so promotes rain. Hence they are taken for oblation in the winnowing basket made of dried straw, reeds or bamboo. Rice and barley are collected in the basket of straws, reeds or bamboo since both rice and straw/reeds/bamboo are rain promotor. They are not anti to each other.'

Thus in view of the facts and circumstances discussed above, it can unhesitatingly be maintained that rainmaking and prevention of rain (to be discussed in the next chapter) is a matter of coordination between the six climatic conditions and six types of vegetations to be offered in the *Yaj¤iya* fire depending upon the careful selection and handling of *Èhuti dravya* (vegetation) in respect of the prevaling climatic condition.

Notes and References

1. *VS.* 18.9.

2. *Ibid.* 22.22.

3. *Ibid.* 2.16.

4. *AV.* 5.24.5.
5. *VS.* 33.3.
6. *Nir.* 2.16.
7. *Ś. Br.* 5.3.5.17.
8. *VS.* 6.29.
9. *VS.* 2.16.
10. *Ibid.*
11. 1.35.4.
12. *RV.* 1.35.4.
13. 4.12.2.
14. 1.68.
15. *VS.* 6.11.
16. *Vāyupūrāṇa,* 51.15.
17. *Ibid.* 51.25.
18. 2.4.8.
19. *AV.* 4.12.7.
20. *Ibid.* 4.12.8.
21. *Ibid.* 4.12.1.
22. *Ibid.* 6.22.3.
23. *VS.* 2.16.
24. *Ś. Br.* 1.1.4.19.
25. *Ibid.* 1.1.4.20.

16

Prevention of Rain

The idea of prevention of rain though is quite old, one could not strive to speculate or conduct experiments in this direction. In modern times, it finds a practical shape only at the hands of Ram Narain Arya who has conducted over 50 experiments at various places and various occasions to ward off the rain from 1973 onward. Like that of rainmaking, rain can also be stopped as and when required. Whereas, the coordination of *Mitra* and *Varuṇa,* the main agents of rainmaking, in a particular ratio, help stimulate rains, the impairment of this coordination causes the rain stop. This coordination of elements in a particular ratio can be marred by inducing the excess quantity of either of two agents into the atmosphere. This further depends on the prevalent weather conditions, which of either two agents is to be introduced in the excess quantity by offering the concerned oblations *(āhutis)* to the *Yajñīya* fire. Here, it can unhesitatingly be said that the viewers will get thrilled to see the *Yajñīya* place to be completely out of reach of the falling rain.

In fact, prevention of rain is quite an opposite phenomenon to that of rainmaking in a sense that the distribution of rainfall, is a subject to the movements of clouds which is further controlled by the speed and direction of uncontrolled winds. Hence, the rain can not be made to fall in as much precision as it can be made to stop by a rainmaker, and the result is that the quantity of occurrence of rainfall may differ from place to place. On the other hand, the phenomenon of the prevention of rain remains under the perfect control of an individual and the local area (*Yajñīya* place) can be saved completely from being lashed out even by torrential rain between half an hour of starting the *Yajña*.

The principles of prevention of rain are just the reverse to the principles of rainmaking. So, further elaboration in this regard is not required.

17

Appendix-1

(Prevention of Diseases)

Actually the diseases of human beings, animals and plants are caused by the impairment of the equilibrium of *vāta, pita* and *kapha,* i.e. air fire and water respectively in the body. The treatment of *vāta, pita* and *kapha* is often done by administering various suitable *auṣadhis* (herbs). Life of plants, animals and human beings is sustained by the food they intake and the air they inhale. So, either way, their treatment is possible. Through the oral intake of *auṣ adhis* as foods as well as through inhalation of the diffused molecules/atoms of the concerned *auṣadhis* in the air. According to the expanding characteristic of the air, as stated earlier, a solid or liquid substance when treated with fire converts into its gaseous state thus facilitating the diffusion of its molecules/atoms with that of the air. The molecules of herbs thus diffused with the air when inhaled produce an effect similar to that of an *auṣadhi* taken orally. Moreover, the administration of *auṣadhi* through *Yajña* has far more and larger effect than its oral administration. For instance, the oral intake of *auṣadhi* may affect only the patient who has taken it, but the administration of the same through *Yajña* may affect a large number of patients on account of its diffusion with the air. This method is useful to take preventive steps to check effectively the spread of epidemics, etc. It may also be used for the cure of people at large scale.

18

Appendix-II

(Pollution Control)

Air pollution can also be checked with the help of *Yajña* since the *Yajña* is helpful in inducing rain and blowing winds in the required direction and with required velocity. Thus rainmaking and strong sway of winds at times may help reduce pollution in densely populated cities and industrial areas of the country. In addition, certain herbs such as mustard, rape-seed, *guggul* (bdellium olibanum) five type of *kaṇṭakāris* (solanum jacquint), both basils, mint, *arka* (swallow wart or asclepias gigantia), *apāmārga, saṅkhapuṣpī,* camphor and *ghee* have been identified by Indian seers as purifiers. If their oblations are offered to fire, their smoke acts as a purifying agent of the atmosphere. As such everybody can perform *Yajña* with the help of above-named material in order to purify the surrounding environs.

19

Appendix-III

Agniṣomīya Paśuyāga

(An Operation for Rainmaking)

Vedas are the first and foremost record of the great advance made by humanity. Vedic Ṛṣis explored the creation undergoing the journey from metaphysical through physical to astrophysical. These were known as three aspects of creation: *ādhyātmika*[1] (metaphysical), *ādhibhautika*[2] (physical) and *ādhidaivika* (astrophysical). During their explorations, they found that all three aspects are interdependent. Physical is based on the Astrophysical aspect and Astrophysical on metaphysical, the metaphysical aspect being the primary source of evolution. The other way round, it can be maintained that from metaphysical evolves Astrophysical and from Astrophysical evolves the physical one. Similarly, during the time of dissolution physical dissolves into astrophysical and astrophysical finally, dissolves into metaphysical. Vedic scholiasts tried to define all the three inter-dependent aspects through equivalence. That is why, the terminology and terms applied by them have the equivalence in the fields of *adhyātma, adhidaivata* and *adhibhūta.* The Vedic visionaries who visualised the laws of *parā* and *aparā* nature beyond time and space applied various methods to define the laws of *parā* nature (metaphysical), *aparā* nature (astrophysical and physical sciences). All these methods were figurative signifying *adhyātma* (metaphysical), *adhidaivata* (astrophysical) and *adhibhūta* (physical aspect). *Yājñas* were not the ends but they were means to elucidate and explain the physical, astrophysical and metaphysical intents of the Vedas. In fact, metaphysical was the primary intent of the Vedas, astrophysical and physical intents being dealt with secondarily. Similarly, Brahmanic ritualism was subjected to *adhyātma* (metaphysical) as their primary significance and others such as *adhidaivata* (astrophysical) and *adhibhūta* (physical) are dealt with as the secondary significance. A close perusal of the *Brāhmaṇas* confirms this fact. The *adhyātmika* intent of the *yajñas* has been referred to 100 times therein, while astrophysical and physical intents have been referred to 60 times and 9 times respectively (for detail see RPA, 1991, 59-60).

In addition to this, hosts of instances may be cited where the *yajñas* can be shown to have been intended by authors of the *Brāhmaṇas* for representing the knowledge of *adhyātma.* For instance *Ś.Br.* 10.2.5.1&2 narrates the aim of *Agnicayana* ceremony as to *ātmasanskāra* or self-purification:

tathaivaitad yajamāna etāḥ puraḥ prapadyābhaye' nāṣṭrā etamātmānaṁ sanskurute.

Confer Sāyana's commentary here:

tathā'yaṁ yajamāno'pi upasadormadhye agnicayanenātmānam sanskurute.

At another place Ś. Br. describes all the *yajñas* aiming at *ātmasampādana* or self-accomplishment, e.g.

sarvairhi yajñairātmānam sampannam vide (10.2.615)

At yet another place Ś. Br. proclaims *Agnihotra* as the expounder of *Ātman* or *Brahman, i.e.,* Supreme self, e.g.

Śauceyo ha prācinayogya uddālakam āruṇimājagāṁ brahmodyaṁ agnihotraṁ vividisiṣyāmi iti. (11.5.3.1)

We come across a reference in *Ś.Br.* where *Darśa-pūrṇamāsa* and *Cāturmāsya* sacrifices are depicted as having twofold objects of their performance, *viz. ātmayājītva* (i.e. metaphysical or manifestations of self) and *devayājītva* (astrophysical), as in:

prajāpatiraha cāturmāsyairātmānam vidadhe (11.5.2.1)

Sāyana's commentary is noteworthy here. According to him,

> *yathā darśapurṇamāsayājina devayājitvaṁ ceti dvaividhyaṁ darśitaṁ evaṁ cāturmāsya yājino'pi tathātvaṁ darśayituṁ śarīrāvayavakalpanam ākhyikayā racayati.*

Thereupon as regards the question as to who is superior between *devayājī* and *ātmayājī Ś.Br.* clarifies :

> *ātmayājī śreyān devayāji iti, ātmayājīti sa ha ātmayajīno vededaṁ me'nenāṅgam sanskrīyate idaṁ me'nenāṅgam upadhīyata iti. (11.2.6.23)*

That is *'ātmayājī* is superior between *ātmayājī* and *devayājī. Ātmayājī* is he who knows while performing *yajña* which of the parts of his body is being consecrated or augmented by a particular action of *yajña.*'

Thus the internal evidence of the *Brāhmaṇas* gives support to the view that physical, astrophysical and metaphysical meanings in order of preference were indicated by the ceremonial rituals of the *Brāhmaṇas.*

Agniṣomīya paśuyāga (ASPY) being one in the series also represents the metaphysical, astrophysical and physical aspects. In a metaphysical sense, it is the charging of an individuated consciousness with knowledge. *Paśu* here means a curious student or an individual (individuated consciousness) who is able to perceive the world around him but is devoid of knowledge or cognition of the same. *Paśu saṁjñapana* means to make the individuated consciousness surcharged with knowledge. Complete surcharging leads to the universalization of consciousness or *mokṣa* from the physical body. *Agni* means a charge of knowledge or says a Guru or a teacher and *Soma* is a good conductor of charge or say a student worthy of charging with knowledge. A *Soma* is always sacrificed to *Agni* for the accomplishment of the process of surcharging. This subject has been handled in detail by Swami Dayānanda Saraswati in course of rendering his interpretation of *Yajurveda.*[1] Here knowledge is the property of *ātman* or consciousness and until and unless the consciousness is completely surcharged with knowledge, it cannot attain universal-hood or *mokṣa.* The seers had it as :

ṛte jñānāt na muktiḥ.

But here it may be pointed out that the *Agniṣomīya paśuyāga* has its equivalence at astrophysical as well as physical levels also. Physically it represents annihilation process of agneya (-vely charged particle) and somiya (+vely charged particle) matter or particles known in modern science as matter and anti-matter. Astrophysically it represents the method of rainformation which is the main subject to this article.

In fact, when the metaphysical aspect is signified, the *yajñas* assume the character of *Brahmayajña;* while signifying astrophysical aspect they are termed as *Devayajñas.* Thus *Brahmayajña* form of ASPY signifies the metaphysical aspect and *Devayajña* form of ASPY signifies the astrophysical sense. The purpose of *Devayajña* form of ASPY is narrated in the *Yajurveda mantra* as follows:

adbhyas tauṣadhibhyaḥ.... (*VS.* 6.9)

That is, for the sake of waters and vegetation we procure you O *havis.*

The same *mantra* further reads:

anu tvā mātā manyatām anu pitānu bhrātā sagarbhyo'nu sakhā sayūthyaḥ. (6.9)

'Let your parent herbs, sister herbs, and simultaneously born herbs and your friendly herbs permit you to be procured for rain and growth of vegetation.'

During the course of translating this stanza Uvaṭa clarifies the same with a quotation from *Śruti.*

idaṁ hi yadāvarṣatyath auṣadhayo jāyante

'When it rains, herbs grow.'

Thus the main purpose of *Agniṣomīya paśu yāga* in *adhidaivata* sense works out, as is evident from the actual verse of the *VS.* to be rainmaking. This is why the seer says that the *paśu* (vegetation) endowed with the quality of *Agni* and *Soma* is procured for *yajña.*

अग्नीषोमाभ्यां जुष्टं नि युनज्मि ।। 6.9 ।।

agniṣomābhyām juṣṭaṁ niyunajmi (*VS.* 6.9)

And the type of vegetation i.e. vegetation worthy of augmenting the power of *Agni* and *Soma* are consecrated. *Agniṣomābhyām tvājuṣṭaṁ prokṣāmi* (*VS.* 6.9). In fact, for making it rain, *Agni* and *Soma* are coordinated in a particular ratio.

apāṁ ca jyotiṣaśca miśrībhāvakarmaṇo varṣa karma jāyate

(See for detail RPA 1995: 146)

These two elements *Agni* and *Soma* are defined in Ś. Br. as under:

dvayaṁ vā idaṁ na tṛtīyamasti. ārdraṁ caiva śuṣkaṁ ca. yacchuṣkaṁ tadāgneyaṁ yadārdraṁ tat somyaṁ.
(Ś. Br. 1.6.3.23)

'There are only two elements, No third one is there. One is dry, another is wet. Dryness pertains to *Agni* and wetness pertains to *Soma.'*

Here the great debatable term is *paśu.* This *yajña* has primarily been taken by many ancient and modern Vedic scholiasts for animal killing. According to them, including Mahidhara, one of the commentators of *VS.,* in *ASPY* animal is sacrificed. So, the tradition of animal sacrifice is also traced back to the Vedas themselves.

In fact, scholars could not make out the actual intent of *paśu.* So they speculated *paśu* for its conventional sense, i.e., animal. Now we shall try to ascertain the actual meaning of *paśu* intended by the seer on the basis of internal evidence of *mantras* implied in *Agniṣomīya paśuyāga* and their explanation in *Ś.Br.* in course of handling the same subject. *Ś.Br.* (3.8.4.5) defines *paśu* broadly as life essence. This *paśu* is the oblation metaphysical for all devas (natural forces).

prāṇa eva paśuḥ sarvāsāṁ vai devānāṁ haviḥ paśuḥ.

(*Ś.Br.* 3.8.4.5)

A.Br. also points out that *paṣu* is the oblation-material.

havirhi paśuḥ

Now the question arises as to what was the actual nature of *paśu* to be sacrificed as the *havi* for deities in the context of *Devayajña.* Whether it was animals or plants. To clarify it we may quote here the seer of the *Yajurveda.* According to him, *vanaspati* or vegetation is selected for *devayajña.*

तं त्वा जुषामहे देव वनस्पते देवयज्यायै । ।5.42 । ।

taṁ tvā juṣāmahai deva vanaspate devayajñāyai.
(*VS.* 5.42).

'O vegetation, we select you as oblation-material for *devayajña.'*

Sacrificing material in *devayajña* is plants or foodgrains and nothing else; it is proved by the following reference of the *Ś.Br.* (1.2.1.20):

ulūkhala mūsalābhyāṁ dṛṣadupalābhyāṁ havir yajñaṁ ghnanti.

'The sacrificing material for *devayajña* is pounded with the help of *ulūkhala* and *mūsala* (i.e. mortar and wooden pestle or grinding board and muller).'

Thus the act of *puśu's* killing with *ulūkhala* and *mūsala* is, in fact, the preparation of *havi* to be offered to the fire of *yajña.*

ghnanti vā etat paśuṁ yadagnau juhvati.

(*Ś.Br.* 3.8.1.10)

At one another place *Ś.Br.* clearly, states that only vegetation is used for *yajñas.* Men could not have performed *yajña* but for vegetation. It is said that *yajñas* are performed only with vegetation.

vanaspatyo hi yajñiyā. na hi manuṣyā yajñeran yad vanaspatayo na syuḥ. tasmād āha vanaspatir yajñiya iti.

The *Śatpatha* reference further makes it clear. Accordingly, by offering vegetation to *yajñīya* fire, we are not destroying them at all.

na vā etaṁ mṛtyave nayanti yaṁ yajñāya nayanti,
(*Ś.Br.* 3.8.1.10)

They cannot get eliminated by this way. In fact, they become imperishable. They again grow and get back to life (flourish) on the earth.

amṛtam āyur hiraṇyaṁ. tad amṛtaṁ āyuṁṣi pratitiṣṭ hati. tathāta udeti. tathā sañjīvāti. (*Ś.Br.* 3.8.1.10)

Not only the author of the *Śatapatha,* but the seer himself expresses nonetheless the similar view. He had it, as a result of *yajña* through the plants or foodgrains, rainy waters flow on the earth. These rainy waters will relieve you of the sin of cutting and destroying the vegetation since more and more vegetation will, in turn, be able to grow.

इदमापः प्र वहतावद्यं च मलं च यत्। यच्चाभिदुद्रोहानृतं यच्च शेपे अभीरुणम्। आपो मा तस्मादेनसः पवमानश्च मुंचतु।। 6.17।।

idamāpaḥ pravahatāvadyaṁ ca malaṁ ca yat. yaccābhi dudrohānṛtaṁ. yacca śepe abhīruṇaṁ āpo mā tasmādenasaḥ pavamānañca muñcatu. (*VS.* 6.17)

Paśu was (*anna*) foodgrains. It is clearly indicated by *Ś. Br.* at one place as

na ha vā etasmā agre paśavaścakṣamire. yadannam abhaviṣyan yatheyam annaṁ bhūtā. (3.7.3.2)

'The *paśus* were therefore not able to see' in the

beginning since foodgrains were called *paśus.*

Further, the seer tells us as to what sort of vegetation would be worthy of use as an *Agniṣomīya paśu* or material for offering as an oblation to ASPY.

According to the seer, the vegetation that has grown in water or has consumed a lot of water is worth to be the *deva havi* (divine oblation) in *Agniṣomīya paśuyāga.*

अपां पेरुरस्यापो देवी: स्वदन्तु स्वात्तं चित्सद्देवहवि:।

सं ते प्राणेा वातेन गच्छतां सम‌ङ्गानि यजत्रै: सं यज्ञपतिराशिषा।।१०।।

apāṁ perūrasyāpo deviḥ svadantu svāttaṁ citsaddeva haviḥ. saṁ te prāṇo vātena gacchatāṁ sam aṅgāni yajatraiḥ saṁ yajña patirāśiṣā (*VS.* 6.10)

So, it must be made in mind that the *havi* to be used for rainmaking or cloud-seeding should be of *somīya* nature, i.e. it should have grown in the climate of high humidity and heavy rainfall.

Bṛhadāraṇyaka Upaniṣad also reflects an ample good light upon the rain-inducing effect of *somīya āhutis,* such as,

somaṁ rājānaṁ juhvati. tasyā āhūtyai vṛṣṭiḥ saṁbhavati.

(*RPA* 1995: 152)

During our experiments of rainmaking, this fact was found to be proven, since all of our experiments for cloud-seeding and rainmaking were accomplished only with the help of *Somīya āhutis.* This fact has been disclosed in detail by the author in his work Vedic Meteorology, part II, chapter 3.

Only *somīya havis* are not sufficed to induce rain or seed clouds, but they should be committed to the fire of *yajña* in the company of *Ghṛta* or fats obtained from the milk of a cow. The verse had it as:

घृतेनाक्तौ पशूँस्त्रायेथाँ रेवति यजमाने प्रियं धा आ विश।
उरोरन्तरिक्षात्सजूर्देवेन वातेनास्य हविषस्त्मना यज समस्य

तन्वा भव। वर्षो वर्षीयसि यज्ञे यज्ञपतिं धाः स्वाहा देवेभ्यो देवेभ्यः स्वाहा। ।6.11 । ।

ghṛtenāktau paśūṁs trāyethā revati yajamāne priyaṁ dhā āviśa. urorantarikṣātsajurdevena vātenāsya haviṣastinanā yaja somasya tanvā bhava. varṣo yajña-patiṁ dhāḥ svāhā devebhyo devebhyaḥ svāhā. (*VS.* 6.11)

'Let the *somīya āhutis* be soaked in the *Ghṛta.* Only after being soaked, they should reach the *yajamāna.* They should be offered to *yajña* in the presence of air in the open sky. This way they get expanded in volume. since they are born in rain, they will induce rain. 'Let the *yajamāna* offer oblation in the name of the concerned deity or to augment the power of the concerned deity.'

Not only *somīya havis* are soaked in *Ghṛta,* but the oblations of a fairly good amount of *Ghṛta* is also offered in this *yajña.*

घृतस्य कुल्या उप ऋतस्य पथ्या अनु।।१२।।

ghṛtasya kulyā upaṛtasya pathyā anu. (*VS.* 6.12)

In our experiments of rainmaking, we discovered that *Ghee* is also one of the ingredient materials to be offered to the fire of the *yajña* for rainmaking. The seer has visualised this fact thousands of years ago.

After giving a prescription of the offered material consisting of *Ghee* and other *somīya* vegetation, the seer extols the quality of celestial waters or water vapours abiding in mid-sphere and seeks the ability to procure them. For example,

देवीरापः शुद्धा वोढ्व ँ सुपरिविष्टा देवेषु सुपरिविष्टा वयं परिवेष्टारो भूयास्म। ।6.13 । ।

devirāpaḥ śuddhā voḍhvaṁ supariviṣṭā deveṣu supariviṣ ṭā vayaṁ pariveṣṭāro bhūyāsma.

(*VS.* 6.13)

'The celestial waters are pure and exist everywhere in the

upper region. They have found their place among other *devas,* i.e., *Indra* etc. Let us become capable of nabbing them or procuring them.'

In the next verse, each and every part of the *havi* is purified so that it may become worthy to be offered as an oblation.

वाचं ते शुन्धामि प्राणं ते शुन्धामि चक्षुस्ते शुन्धामि । ।6.14 । ।

vācaṁ te śundhāmi prāṇaṁ te śundhāmi, cakṣuste śundhāmi.

(VS. 6.14)

Further, it was prayed that each and every part of the oblation material should be in a sound condition. For example, the seer likes it to be as:

मनस्त आप्यायतां वाक आप्यायतां प्राणस्त आप्यायतां चक्षुस्त आप्यायता ँ श्रोत्रं त आप्यायताम्। यत्ते क्रूरं यदास्थितं तत्त आप्यायतां निष्ट्यायतां तत्ते शुध्यतु शमहोभ्यः ।

manas ta āpyāyatāṁ vāk ta āpyāyatāṁ prāṇas ta āpyāyatāṁ cakṣus ta āpyāyatāṁ śrotraṁ ta āpyāyatāṁ yatte kruraṁ yadāsthitaṁ tat ta āpyāyatāṁ niṣṭyāyatāṁ tat te śudhyatu śamahobhyaḥ. (*VS.* 6.15)

Thus it is obvious from the foregoing that oblation-material should be pure and in a sound condition. There should be no impurity; it should not be in a decayed position. For this purpose, he asks the plant to provide the offering material intact. *oṣadhe trāyasva* (*VS.* 6.15). He further suggests not to destroy the plant completely with the cutting knife or sword. *svadhite mainaṁ hiṁsiḥ.* (6.15)

In the beginning, the seer has advised not to go for impure or mixed material, but in spite of all care duly taken in selecting the *somīya* offering, there is the possibility of some material not being *somīya* and so creating the anti-rain effect. In such conditions, this type of material may necessarily be separated or removed. The seer explains this operation as under:

रक्षसां भागोऽसि निरस्त ँ रक्ष इदमह ँ रक्षोऽभि तिष्ठामीदमह ँ रक्षोऽव बाध इदमह ँ रक्षोऽधमं तमो नयामि।

rakṣasām̐ bhāgo' si nirastam̐ rakṣa idam aham̐ takṣ a'bhitiṣṭhāmidamaham̐ takṣo' vabādham̐ idamaham̐ takṣa' dhamam̐ tamo nayāmi. (*VS.* 6.16)

'You create anti effects. I remove you since you are anti-material. I remove this anti-material. I stop this to be used. I throw it to the heap of rubbish.'

He also discloses the process as to what happens after *somīya āhutis* are offered to fire.

According to him, on being sacrificed the *āhuti* goes to the upper layer of atmosphere along with the air drifted upward on account of the heating effect of *yajñīya* fire.

स्वाहाकृते ऊर्ध्वनभसं मारुतं गच्छतम् । ।6.16 । ।

svāhākṛte ūrdhvam̐ nabhasam̐ mārutam̐ gacchatam̐. (*VS.* 6.16)

In fact, the *āhuti* goes to the sky in the form of its essence or vapours, its carbon part scatters on the earth.

दिवं ते धूमो गच्छतु स्वज्योर्तिः पृथिवीं भस्मनाऽऽपृण स्वाहा । ।6.21 । ।

divam̐ te dhūmo gacchatu svar jyotiḥ pṛthivī bhasmanā pṛṇa svāhā. (*VS.* 6.21)

Thus the *āhutis* reach their concerned deities in the form of smoke.

At the end, the purpose of *Agniṣomīya paśuyāga* is further explained by the seer's prayer to *Varuṇa* for releasing waters and herbs from their respective places of origin, instead of destroying them. He asks for a pardon because though the herbs were not worth destroying, but they had destroyed them by sacrificing into the fire. He prays further, 'Let these waters and herbs be useful for us, let them be harmful to those who are harmful to us.' The stanza reads as follows:

माऽपो मौषधीर्हिं ँ सीर्धाम्नो धाम्नो राजँस्ततो वरुण नो मुंच। यदा हुरघ्न्या इति वरुणेति शपामहे ततो वरुण नो मुंच। सुमित्रिया न आप ओषधयः सन्तु दुर्मित्रियास्तस्मै सन्तु योऽस्मान्द्वेष्टि यं च वयं द्विष्मः। ।6.22। ।

māpo mauṣadhirhiṁsī dhāmno dhāmno rājanstato varuṇa no muñca yadāhuraghnyā iti varuṇeti śapāmahe tato varuṇa no muñca sumitriyā na āpa oṣadhayaḥ santu durmitriyastasmai santu yo asmāndveṣṭi yaṁ ca vayaṁ dviṣmaḥ. (*VS.* 6.22)

'Let the waters and herbs be not destroyed. Let them flourish at the respective places of their origin. Let the *Varuṇa* release them for us from the respective places of their origin. They are said not worth destroying and we are blamed to have destroyed them. So please O *Varuṇa,* excuse us. Let both the waters and herbs be friendly (useful) with us; let them be unfriendly with those who are harmful to us and who (bacteria) we want to destroy.'

This stanza marks the end of *Agniṣomīya paśuyāga.* Thus it is crystal clear from the aforementioned discussion that the aim of *Agniṣomīya paśuyāga* is rainmaking or cloud-seeding so far as its astrophysical intent is concerned.

This *yāga* has nothing to do with animal sacrifice as proposed by ancient commentators of *Yajurveda,* like Uvaṭa, Mahīdhara and several modern occidental and oriental scholars. In fact, the use of the word *paśu* creates this type of misconception in the mind of the readers. They are misled by *puśu* taking it to mean animal, but the actual intent of *paśu* here is the vegetation to be offered as an oblation to the fire.

havirhi paśuḥ

Moreover, one should not forget that the seer of *Yajurveda* himself declares that *vanaspati* or vegetation is used for *devayajña.*

तं त्वा जषामेहे देव वनस्पते देवयज्यायै। ।5.42। ।

taṁ tvā juṣāmahe deva vanaspate deva yajñāyai.(*VS.* 5.42)

So far as the various parts to be offered to the fire of *yajña* are concerned, they fit suitably more in the context of *ādhyātmika* sense than in astronomical sense.

REFERENCES

Ravi Prakash Arya (RPA), *Researches into Vedic and Linguistic Studies,* Grantha Bharati Prakashan, Delhi, 1991.

Uvaṭa and Mahīdhara, *Vājasaneyi Saṁhitā (VS). Motilal Banarsidass, Varanasi.*

Śatapatha Brāhmaṇa (Ś.Br.), Rashtriya Sanskrit Sansthan, Delhi, 1990.

20

Appendix-IV

Somayāga-The Process of Rainformation

As it has been repeatedly stated by the present author that various Vedic *Śrauta yajñas* starting from *Agnyādhāna* to the *yajñas* running into 1000 years explain/represent, in an astrophysical sense, the various phases of the process of physical or astrophysical creation that is going to last till 1000 *Caturyugas* or *Mahāyugas* i.e. 311 billion years. In a spiritual sense, the same *yajñas* represent the process of emancipation of souls in this universe. For instance, in a spiritual sense, *Agnyādhyāna* explains/represents the existence of *ātmans* and in an astrophysical sense, the same represents the first origin of *āgneya* or fire element on the earth. In a physical sense, it represents the origin of first charged particles from the energy. *Agnihotra* in a spiritual sense explains/represents the oblations of *soma* or *prakṛti* (matter) element on the *puruṣa*/*agni* (consciousness) element. In the astrophysical sense, it represents the change of day and night. In a physical sense, it represents the creation of matter from energy. Similarly *Darśapūrṇamāsa* explains/ represents the change of fortnights, *Cāturmāsya* explains/represents the seasonal changes and *Gavāmayana* explains/represents changes of winter and summer solstices i.e. *Uttarāyaṇa* and *Dakṣiṇāyana* on the earth, similarly *Agni-somīya-paśuyāga* explains the operation for rainmaking whereas *Somayāga* explains/represents the process/phenomenon of rain formation on the earth.

Somayāga: To start with, it is necessary to inform you that there are two types of *yāgas*, *Śrauta yāgas* and *Smārta yāgas*. *Smārta yāgas* have been described in *Gṛhya Sūtras* in the name of Various *Sanskāras*. They are called *Pākasanstha*, because they end with some preparations. *Śrauta yāgas* have been described in *Śrauta Sūtras*. They are called *Havisanstha* and *somasanstha*. *Havisanstha* are those that end with oblations of *puroḍāśa* (vegetation in their ripened form) into the fire of *Yajña* (annihilation of particles and anti-particles in the physical sense). *Somasanstha* are those that end with oblations of *soma* (essence or juice of vegetation) into the fire. *Puroḍāśa* represents vegetation, herbs or particles and *soma* represents the juice of vegetation (electrical charge in the physical sense). In fact, *Somasanstha yāgas* represent the conversion of energy into matter in the universe, whereas *Havisanstha yāgas* represent the conversion of matter into energy or annihilation of matter and anti-matter in the physical sense. It represents some operation carried out by human beings to bring about desired changes in the environment or say for modifying the environment according to the needs and requirements of animate beings on the earth. *Havisanstha yāgas* are of seven types *viz. Agnyādheya*, *Agnihotra*, *Darśa*, *Pauraṇamāsa*, *Āgrayaṇa*, *Cāturmāsya* and *Paśubandha*. Thus all *Paśuyāgas* are included into *Havisanstha Somayāgas*. Here it may also be known that in the *Havisanstha Somayāgas*, the term *Paśu* doesn't signify a conventional sense of animal, but a '*havī*' or 'oblation' prepared especially for some targeted *devatā* , in general, is called *Paśu*. In a physical sense, it represents a charged particle. Similarly *Sautrāmaṇī Iṣṭi* is also a form of *Havisanstha Somayāga*. *Soma-sanstha Somayāgas* are also of seven types. They are known as *Agniṣṭoma*, *Atyagniṣṭoma*, *Ukthya*, *Śoḍasi*, *Vājapeya*, *Atirātra*, *Āptoryāma* and *Aṣṭaka*.

Somayāga is started with *agniṣṭoma yāga* which is accomplished within six days. On the First-day *dikṣaṇīyeṣṭi* is performed. This indicates the *dikṣā* or initiation of *Yajamāna* and his wife for the *Somayāga*. Since the *Somayāga,* in an astrophysical sense, represents the process of rain formation on the earth, the initiation of *yajamāna* and his wife is essential. The process of rain formation is accomplished with the help of *agni* and *soma*.

> *apāṁ ca jyotiṣaśca miśrībhāvakarmaṇo varṣa karma jāyate.* (RPA:146).

These two elements *Agni* and *Soma* are defined in *Ś. Br.* (1.6.3.23) as under:

> *dvayaṁ vā idaṁ na tṛtīyamasti. ārdraṁ caiva śuṣkaṁ ca. yacchuṣkaṁ tadāgneyaṁ yadārdraṁ tat somyaṁ.*

'There are only two elements, No third one is there. One is dry, another is wet. Dryness pertains to *Agni* and wetness pertains to *Soma.*'

Here *yajamāna* represents *agni* and his wife represents *soma*, so initiation of both is essential for this *yāga* to take place.

The next three days are devoted to the observance of *Iṣṭ is* and *Pravargay*. Fifth-day *soma* is procured and *Somayāga* is accomplished in three phases as *prātaḥ savana* (in the morning hours), *mādhyandina savana* (during noon) and *sāyaṁ savana* (evening hours). These three phases represent the three phases of an year, each phase consisting of 4 months (*Cāturmāsya*), the first phase being the rainy season, the second phase being winter season and the third phase being summer season. On the Sixth-day closing is marked with the *avabhṛtha* rite. In this *yāga*, which accomplishes in three phases/seasons of 4 months each (represented respectively by three days i.e. 4th, 5th and 6th day) three types of *paśus* are procured in each phase. In the first phase representing the rainy season consisting of 4 months (*Cāturmāsya*) represented by the fourth day of *yāga*, *agniṣ omīya paśu* (that is rain-bearing monsoon formed of *āgneya* and *somīya* elements) is procured.

On the fifth day representing the second phase of winter season consisting of 4 months (*Cāturmāsya*), *savanīya paśu* (seeded cloud that which delivers rain in the rainy season) is procured.

On the sixth day representing the third phase of summer season consisting of 4 months (*Cāturmāsya*), *maitrāvāruṇī vāśā anubandhyā gau* (vaporisation caused by *Mitra* and *Varuṇa* which is not capable of delivering rain) is procured.

The *Somayāga* is accomplished with the following ceremonial acts.

1. ***Pravargya*:** In *pravargya* ceremony, *ghee* is heated into an earthen pot and the *ghee* thus heated is mixed with the milk of cow or goat. The heating of *ghee* in this ceremony symbolises the radiation heating of earth in the summer season and pouring of milk symbolises the evaporation of waters that takes place due to radiation heating.

2. **Purchase of *Soma*:** The next ceremony is to purchase the *Soma* from *Ekahāyanī Aruṇā gau.* Here *eka* (one) *hāyana* represents one year and *aruṇā gau* represents red-hued lightning of the clouds. This ceremony indicates that *soma* (rainy waters) are received from the red-hued lightning occurring in the rainy season after one year. The different hue lightning heralds different met phenomenon. For example:

vātāya kapilā vidyudātpāyātilohini kṛṣṇā sarvanāśāya durbhikṣāya sitā bhaveta

That is *kapilā* (yellow) coloured lightning signals storms and cyclones, violet coloured calls for very heat. Black hue signals destruction and white coloured points out the famine.

3. ***Agniṣomīya Paśu*:** *Agniṣomīya paśu* is nothing but the maritime winds (blowing from sea to continents) are called *somīya paśu* and continentals winds (blowing from dry reasons) are called as *āgneya paśu*. In fact, air/the wind, fire and the sun have been described as *paśu* in the *Yajurveda*. For example, *Yajurveda* (23.17) says:

agni paśur āsīt, tenāyajanta, vāyu paśur āsīt tenāyajanta, sūryaḥ paśur āsīt tenāyajant.

That is in the process of creation (*adhidaivika yajña*) fire acted as *paśu* (animal) to be offered as an oblation. Similarly, wind/air acted as an animal which was also sacrificed and the sun also acted as *paśu* and so was also sacrificed as an oblation. This means to say that the physical creation evolved out of the fire, air and sun.

Thus *Agniṣomīya paśu* refers to monsoon winds or other winds capable of bringing rain.

4. ***Havirdhāna maṇḍapa*:** *Havirdhāna maṇḍapa* signifies the place where the *havis* are escaped. In the astronomical context of *Somayāga*, mid sphere or *antarikṣa* acts as the *havirdhāna maṇḍapa* where the evaporated waters (*soma*) are located. *Soma* (evaporated waters) are the oblations of *Somayāga*. Here it may be informed that *pavamāna soma* has been a deity in the Vedas which signify rainy waters or evaporated waters. These *somas* are called as *pavamāna* as they are distilled or purified waters. The ancient Vedic scholar Yāska reads the term *havi* in the names of clouds. Accordingly, in *Somayāga*, evaporated waters condensed into clouds act as *havi*.

5. ***Havirdhāna Śakaṭa* (Cart carrying *havis*):** In *Somayāga*, there are two *havirdhāna śakaṭas* placed in opposite directions of south and north. These two carts carrying *havis* (oblations of *soma* for *somayāga*) are nothing but the clouds floating in opposite directions carrying the evaporated waters. The opposite flow of these clouds makes them oppositely charged. Here it may be recalled that precipitation takes place with the discharging of oppositely charged clouds in the mid sphere. To explain the above phenomenon of rain formation two carts bearing *havis* are placed in opposite directions in *Somayāga*.

6. ***Soma Abhiṣava* (Delivery of rain):** In *Somayāga*, *soma abhiṣava* (*soma* extraction) is enacted with the help of two stones (mortar and pestle) called *grāvā*. Vegetation or herb is crushed with the help of two *grāvā* to extract *soma*. This process of crushing with the help of two *grāvās* (stones) is followed by a sound. In the Vedic tradition, *grāvā* signifies cloud and not stone. This is clear from *Nighaṇṭu*. Yāska, in

his *Nighaṇṭu*, reads the terms like *grāvā*, *adri* etc. that denote stone in the conventional sense to signify clouds. Thus the act of crushing of *soma* with the help of two *grāvā* in the *Somayāga* signifies the discharging of two oppositely charged clouds contact of which produces a short circuit and unleashes a high amplitude current called lightning followed by acoustic shock waves which are represented here in *Somayāga* by the sound produced by two *grāvās*. The following *mantra* of *Parjanya Sūkta* of *Ṛgveda* (5.83.2) describes the power of such shock waves as *viśvaṁ vibhāya bhuvanaṁ mahāvadhāt*-the entire universe got frightened with the shocking sound caused by the contact of oppositely charged clouds. This process of discharging between oppositely charged clouds leads to the precipitation of rain known as *soma* or *Pavamāna soma* in the Vedas. The same phenomenon of precipitation of rain has been described very amicably in the *Ṛgveda* (1.32.11) as:

dāsapatnīr ahigopā atiṣṭhan
niruddhāḥ āpaḥ paṇineva gāvaḥ.
apāṁ bilam apihitam yadāsīt
vṛtraṁ jaghanvān apa tad vavāra.

That is the waters hidden into the clouds were allowed to fall in the form of rainy waters due to discharging of oppositely charged clouds.

7. Placement of *Adhiṣavaṇaphalaka* (Wooden board): The mortar (stone) upon which *soma* extraction is done is supported by a wooden board called *adhiṣavaṇa phalaka*. This wooden board is called *adhiṣavaṇa* because of the etymology '*adhiṣūyate asmin iti adhiṣavaṇa*. That is, it's being the basis of *soma* extraction. The wooden board represents the layer of the troposphere which is the basis of cloud formation and precipitation of rain.

8. Digging of *Uparavas* (pits): *Uparavas* are a sort of pit that is dug underneath the *havirdhāna śakaṭa* (*havi* bearing cart). They are four in number. They are dug in a way that they appear separated externally but remain one internally. They are covered with two wooden boards. This all symbolises the four different months making one single *ṛtu*/season of rain. The covering by two wooden boards symbolises the covering of four months rainy season by two types of clouds (-vely and +vely charged clouds).

9. Three *Savanas*: The *Somayāga* is accomplished in three *savanas, viz. prātaḥ* (morning) *savana*, *mādhyandin* (noon) *savana* and *sāyaṁ* (evening) *savana.* These three *savanas* denote three seasons of the year, i.e. rainy, winter and summer. *Prātaḥ savana* denotes rainy season, *Mādhyandin savana* denotes winter season and *sāyaṁ savana* denotes summer season. Since *somayāga* gives an account of the distribution of rain throughout the year divided into three seasons of four months each.

10. Distribution of *Soma* in various *savanas*: The total amount of *soma* to be extracted is divided into two parts: larger one and smaller one. The Larger amount is extracted in the morning *savana* and the smaller amount is extracted in the noon *savana.* The pulp of both the extraction is again crushed in the evening *savana* for extraction of *soma.* This process indicates that precipitation takes place in large amount on the earth in the rainy season, which reduces in the winter season. It hardly takes place in the summer season. Thus *Somayāga* points out the distribution of rainy water (*Pavamāna soma*) on the earth throughout the year divided into three seasons of four months each.

11. *Savanīya Paśu* (Animal): *Savanīya paśu* is nothing but the seeded cloud. Seeded cloud is called *savanīya* because that becomes worthy of inducing rain. The tallow of the *Savanīya paśu* is sacrificed in the morning *savana.* Tallow of *Savanīya paśu* (cloud) is nothing but the precipitation. The morning *savana* symbolises, as indicated earlier, rainy season. Through this act of sacrificing of tallow of *savanīya paśu*, the seer wants to tell that the rain is induced from clouds in the rainy season. *Puroḍāśa* is sacrificed in the noon hour (*mādhyandin savana*). *Puroḍāśa* symbolises hailstorms. Hailstorms take place in the noon hour i.e. winter season. In the evening hour, the body parts of *savanīya paśu* are sacrificed. The body parts symbolise the remaining parts of clouds that yield rain in the summer season.

12. *Anubandhyā Maitrāvaruṇī Vaśā*: On the Sixth day or the last day *Maitrāvaruṇī vaśā* is sacrificed in the *Yajña*. Here it may be pointed out that most of the scholar takes *vaśā* for the meaning of 'barren cow'. No doubt, *vaśā* stands for 'barren', but it is absurd to suffix cow to the expression *vaśā*. Since nowhere *vaśā* has been attributed to *gau* or cow. In this *yāga* also *maitrāvaruṇī vaśā* is mentioned. *Mitra* and *Varuṇa* are the rainmaking agents. The Vedic seer proclaims, *mitrāvaruṇau tvā vṛṣṭyāvatām.* (VS. 2.16) 'May the *Mitra* and *Varuṇa* bring rain for you. *Mitrāvaruṇau vṛṣṭ yādhipati tau māvatām.* (*AV.* 5.24.5) 'May the rainmaking agents, *Mitra* and *Varuṇa*, protect you.

In fact, the coordination of both the elements was considered necessary for inducing rain and the proposed coordination could easily be effected with the help of *Yajña.*

yajñā no mitrāvaruṇā yajñā devaṁ ṛtaṁ bṛhat.

'*Mitra* and *Varuṇa*, the main agents of rain should be co-ordinated with the help of *Yajña* for the precipitation of rain.'

We have read above that on the *savanīya* day (fifth day) the *savanīya paśu* or seeded cloud was sacrificed for inducing rain. After the rainy season, seeded clouds no longer exist and the existing clouds are not able to precipitate. Since clouds are formed of *Mitra* and *Varuṇa* elements. Thus the remaining *maitrāvāruṇī* elements (clouds) are known as *vaśā* or barren ones since the clouds are not able to precipitate in the winter season. Unravelling the secrets of *maitrāvaruṇī vaśā* (clouds composed of *Mitra* and *Varuṇa* elements but not able to precipitate for want of required amount of evaporated waters) *Śatpatha Brāhmaṇa* (4.5.1.9) says:

atha yadā na kaścana rasaḥ paryaśiṣyat tat eṣā maitrāvaruṇī vaśā samabhavat. tasmād eṣā na prajāyate. rasāddhi retaḥ sambhavati retasaḥ paśavaḥ.

'That is when no water drops are left there, then the clouds are known as *maitrāvaruṇī vaśā*. Since those clouds (*maitrāvaruṇī vaśā*) are not able to precipitate. The rain or rainy water (*rasa*) causes the production of *retas* (seaman) and living beings are born of *retas*.'

Finally, the *Śatapatha Brāhmaṇa* (4.5.1.9) illustrates the reason as to why this *vaśā* is sacrificed at the end of the *yāga*. Accordingly :

tad yad antataḥ samabhavat. tasmādantaṁ yajñsya anuvartate.

'Because the *maitrāvaruṇī vaśā* is left at the end of *yajña* (phenomenon) of rain formation, that is why here also in *Somayāga* the *maitrāvaruṇī vaśā* is sacrificed at the end.'

Hope now the readers/scholars are able to know the mystery of *Somayāga*. We all must bring in mind that the Vedic knowledge cannot so easily be comprehended. It's not a laymen's literature. To understand the Vedas, it is essential to understand the cultural background of Vedas and the intricacies of the grammar and style of the Vedic Language. Unravelling this mystery, the *Brāhmaṇakāras* say,

prokṣapriyā iva hi devāḥ pratyakṣa dviṣaḥ,

I.e. Vedic seers like to describe the knowledge in secondary (*prokṣa*) sense and they have aversion from relating the mystery in primary (*pratyakṣa*) sense.

Later on, this method of describing the matter in a secondary sense (*taddhitārtha*) was given a grammatical colouring and secondary suffixes were used with the original words to convey the secondary sense. This intricacy of Vedic language has been unfolded by Yāska, an ancient Vedic scholar, illustrating an example. At Nirukta (2.1) in the context of the Vedic expression '*gau*' Yāska says:

gaur iti pṛthivyā nāmadheyam. yad dūraṅgatā bhavat.................athāpyasyāṁ tāddhitena kṛtsnavat

> *nigamā bhavanti 'gobhiḥ śriṇīta matsaram'. aṁśu duhanto'dhyāsate gavi iti adhiṣavaṇacarmaṇaḥ.*
>
> '*Gau* is the name of the earth in astronomical (*adhidaivika*) sense since it goes far away while orbiting the sun. This *gau* is also used in *laukika* (conventional) sense to mean *aukika gau* or 'cow' since the cow also uses to go far off places while glazing. The expression '*gau*', in the Vedas, while denoting '*laukika gau*' or cow implies secondary sense. For instance, in the expression '*gobhiḥ śriṇīta matsaram*' 'go' has been used in the secondary sense. Its primary (*pratyakṣa*) meaning being 'cow' and secondary (*prokṣa*) meaning is 'product of cow' Here the expression *gau* signifies 'product of cow' or say *gopayobhiḥ* (cow milk) and not 'go' 'cow' herself. So the meaning of the above phrase will be 'Cook *soma'* (herbs/vegetables) with the cow milk. Similarly, the phrase '*adhyāste gavī*' (literally meaning 'sits on the cow') also do not signify the primary sense 'sits on the cow', but the very meaning of the above phrase will be 'sits on the leather of cow'.

Thus having illustrated these two example Yāska there itself clarifies that the names of other animals also create similar doubts in the Vedas and allied literature.

evam anyeṣām api sattvānāṁ sandehāḥ vidyante.

Yāska here clearly mentions that when Vedic expressions apply to *laukika* (conventional) things or animals etc. they signify secondary sense and not the primary sense as generally held by the scholars. But when they apply to *ādhyātmika* (spiritual) objects like mind, spirit, God etc. or astronomical objects like stars, planets, they are used in primary meaning. For example, the expression '*gau*' if applies to spiritual objects like sense-organs and to astronomical objects like the earth is always used in the primary meaning, but if the same expression '*gau*' applies to *laukika* or an earthly object like 'cow' is mostly used in the secondary purport. Here I would quote a very interesting example quoted in *Śatpatha Brāhmaṇa* (3.4.1.2) in the context of entertaining a guest. This phrase has also appeared in several *Gṛhya Sūtras* as quoted by Jha (51, footnote 72). According to the reference of *Ś.Br.* (3.4.1.2) the deities are entertained with the help of *havis* just as the guests are entertained with the help of grains produced by Bulls and the milk produced by a goat. Here the *Brāhmaṇakāra* says:

> *yathā rājñe vā brāhmaṇāya vā mahokṣaṁ vā mahājaṁ vā pacet tadāha mānuṣaṁ havir devānām evamasmā etadātithyaṁ karoti.*

> Just as a king or a renowned scholar is entertained with the recipes of grains produced by a strong bull or of the milk produced by a healthy goat respectively. This was the method of entertaining human guests. Similarly, the astronomical guests or natural powers called *devas* are entertained by the oblations offered to the *yajñīya* fire.

Here several scholars translate *mahokṣaṁ* as a big bull and *mahājam* as big goat. Similarly, the words *mahokṣam* and *mahājam* occurred in other *Gṛhyasūtras* are also taken to mean as big bulls and big goats. The translators forget that these expressions are not to be taken in their primary senses, but then in the secondary sense (*taddhitārtha*) and accordingly the *mahokṣa* and *mahāja* will signify recipes made of the grains produced by big bulls and milk produced by big healthy goats. Following the similar notion prevalent about the Vedic language, it would be highly objectionable if a scholar without knowing the actual tendency and peculiarity of the Vedic language uses such phrases in a primary sense when they apply to laukika (conventional) objects. Such a scholar not only supports the inhumane killing of innocent animals but is also guilty of killing the Vedas and actually intended sense of the Vedas.

So with this detailed discussion held above, the author of the present lines thinks that some human sense would prevail upon so-called historians like D. N. Jha and others who are all out to uproot the glorious past of not only of India but the entire earth inhabited by Vedic people.

References

1. *Atharvaveda*, ed. Devi Chand, Munshi Ram Manohar Lal, Delhi, 1982

2. Arya, Ravi Prakash (RPA), *Vedic Meteorology*, Delhi, 1995,

3. Jha, D.N. *The Myth of the Holy Cow*, Verso, London, 2002

4. *Nirukta* of Yāska, ed. with Sanskrit translation, Brahma Muni Parivrajaka. ।

5. *Śatpatha Brāhmaṇa*, with Sayana Bhaṣya, Nag Publication, Delhi.

Select Bibliography

Aeolus: *Meteorology,* Eng. U. Press Ltd., 1952.

Avadha Behari Tripathi: *Bṛhat Saṁhitā* of Varāhamihira, Varanasi, 1968.

Bellikoth Rama: *Jaiminīya Upaniṣad Brāhmana,* K.

Chandra Sharma: Sanskrit Vidyāpiṭha Tirupati, 1967.

Bhagavaddatta: *Ṛgveda par Vyākhyā,* Dayānanda College, Lahore, 1920.

Brahmadatta Jijñāsu: *Dayananda's Yājuṣa Bhāṣya Vivaraṇa.*

Brahma Muni Parivrājaka: *Nirukta Sammarśah,* Sanskrit Tr. of *Nirukta,* Ajmer, 1996.

Dayananda Saraswati: *Yajurveda Bhāṣya,* Ajmer.

Dayananda Saraswati: *Ṛgbhāṣya,* Ajmer.

Dayananda Saraswati: *Ṛgvedādibhāṣvabhūmikū,* de-Yudhisthira Mimansaka, Bahalgarh, Sonepat, 1984.

Eggeling, J.: *Śatapatha Brāhmaṇa,* Oxford, Clarenden Press.

Griffith R.T.H.: *Text of White Yajurveda,* Varanasi, E.J. Lazarus, 1957.

Herbert Riehl: *Introduction to the Atmosphere,* 3rd ed., Mac Graw-Hill, Kogakusha Ltd. 1965.

John Daires: *Saṁkhya Kārikā,* Indological Book House, Delhi, 1957.

Jwāla Prasad Miśra: *Vājasaneyī Saṁhitā,* Bombay, 1969.

Kaegi, A.: *Ṛgveda* Eng. Tr. Boston, Gim, 1886.

Kashinatha Shastri: *Aitreya Brāhmana* with Sāyaṇa

-Bhaṣya, Poona, 1977.

Keith, A.B.: *Ṛgveda Brāhmaṇas, Aitareya* and *Kauṣitaki Brāhmaṇas of Ṛgveda,* Delhi, Motilal, 1971.

Kṣemakaraṇa Trivedi: *Gopatha Brāhmaṇa,* Prayāga, 1920.

Macdonell, A.A.: *Bṛhaddevatā,* Harward Oriental Series, 1904.

Macdonell, A.A.: *Vedic Mythology,* Varanasi Indological Book House, 1963.

Mahavir Sastrin: *Taittirīya Brāhmaṇa* with the commentary of Bhaṭṭabhāskara Miśra, Mysore, 1908-21.

Mason, B.J.: *Clouds, Rain and Rainmaking,* 2nd ed., Cambridge U. Press, 1975.

Max Müller, F.: *Vedas,* Calcutta, Sushil Gupta, 1956.

Max Müller, F.: *Vedic Hymns,* Motilal, Delhi.

Motilal Sharma: *Śatpatha Brāhmaṇa, Vijñāna Bhāṣya,* Jaipur.

Oldenberg, H.: *Sāṅkhāyana Gṛhya Sūtra,* Oxford, 1886.

Pandit, M.P.: *Gems from the Vedas,* Ganesh, 1973.

Raghunandan Sharma: *Vedic Sampatti,* Bombay, 2016.

Raghvir: *Atharvaveda,* S. V. Granthamāla, Lohore, 1936.

Raghvir, Lokesh Chandra: *Jaiminīya Brāhmaṇa,* SVS Nagpur, 1954.

Ram Kumar Rai: *Śaunakīya Bṛhaddevatā,* Varanasi.

Ravi Prakash Arya: *Researches into Vedic and Linguistic Studies,* Indian Foundation for Vedic Science, Delhi, 1991.

Ravi Prakash Arya: *Vedic Science of Weather Modification,* 1993.

Ravi Prakash Arya: *Contrastive Analysis of Vedic and Classical Sanskrit,* Indian Foundation for Vedic

Science, Delhi, 2006.

Ravi Prakash Arya: *Ṛgveda Saṁhitā,* Indian Foundation for Vedic Science, Delhi, 2006

Ravi Prakash Arya: *Yajurveda Saṁhitā,* Indian Foundation for Vedic Science, Delhi, 2006

Ravi Prakash Arya: *Sāmaveda Saṁhitā,* Indian Foundation for Vedic Science, 2006, Delhi.

Ravi Prakash Arya: *Energy in the Vedas,* Amazon Books USA; Indian Foundation for Vedic Science, India 2014.

Ravi Prakash Arya: *Engineering and Technology in Ancient India,* Indian Foundation for Vedic Science, India 2019.

Ravi Prakash Arya: *Rainmaking with the help of Yajña,* Amazon Books, USA, Indian Foundation for Vedic Science, India 2018.

Ravi Prakash Arya: *Indian Chronology to Indian History,* Indian Foundation for Vedic Science, India 2019.

Ravi Prakash Arya: *Introduction to the Vedas,* Amazon Books, USA, Indian Foundation for Vedic Science, India 2016.

Samarpaṇānanda: *Śatapatha Brāhmaṇa* Sāmarpaṇa Bhāṣya. Prabhat Ashram Meerut.

Swami Shridhara Shastrin: *Śāṅkhāyana Araṇyaka,* Anandashram, Poona, 1922.

Shripada Sharma: *Kaṭha Saṁhitā,* Aundh, 1942.

Shripada Sharma: *Kāṭhaka Saṁhitā,* Aundh, 1943.

Shripada Sharma: *Maitrāyaṇī Saṁhitā,* Aundh, 1943.

Shripada Sharma: *Maitrāyaṇī Saṁhitā* (Brāhmaṇa-portion), Aundh, 1943.

Shripada Sharma: *Taittirīya Saṁhitā,* Aundh, 1945.

Uvaṭa & Mahidhara: *Vājasaneyī Saṁhitā,* Varanasi.

Vedapala Sunitha: *Śatapatha ke Daśapatha, Part I & II,* Tilauradham, Rajasthan, 1991-92.

Vedapala Sunitha: *Darśapūrnamāseṣṭirahasya Prakāśa,* Tilaura, Rajasthan, 1991.

Vishva Bandhu: *Ṛgveda-Skandha, Udgitha, Veṅkaṭamādhava Kṛtā Vyākhyā Sahitā,* Hoshiarpur, Punjab.

Vishva Bandhu: *Vedic Word Concordance,* Lahore.